Methylotrophs: Microbiology, Biochemistry, and Genetics

Editor

Ching T. Hou, Ph.D.
Research Associate
Exxon Research Center
Annandale, New Jersey

CRC Press
Taylor & Francis Group
Boca Raton London New York

CRC Press is an imprint of the
Taylor & Francis Group, an informa business

PREFACE

Methylotrophs are microorganisms which utilize compounds containing one or more carbon atoms, but containing no carbon-carbon bonds. They are different from autotrophs which utilize carbon dioxide as the sole carbon source. Although the first methane-oxidizing bacterium was isolated by Sohngen in 1905, the progress of research in this area has been relatively slow. Prior to 1960, only about a half dozen reports were available on these organisms. In the 1960s, most of the research activities on methylotrophs were concentrated in Foster's laboratory in Texas and in Quayle's laboratory in Sheffield, England. Many of the major biochemical problems were solved during this period. Whittenbury and his co-workers classified methylotrophs into several groups on the basis of morphology, fine structure, and type of resting stage formed. In the 1970s and early 1980s, an explosive development of this whole field took place including: more new methylotrophic microorganisms, metabolic intermediates, enzymes, energy transduction, and genetics. New industrial potential of using methane and methanol as cheap and readily available fermentation feedstocks have stimulated research and development in this field. As a reesult many patents were issued in connection with methylotrophs.

The vast amount of literature on methylotrophs has been reviewed on specific subjects at frequent intervals by various investigators. There is one book written by C. Anthony in 1982 covering the biochemistry of methylotrophs. However, there is no single book published which totally devotes itself to and covers all aspects of methylotrophs. This book offers a comprehensive examination of the microbiology, biochemistry, genetics, and applied aspects of methylotrophs. Chapters 1 and 2 are devoted to the microbiology and biochemistry of methylotrophic bacteria and yeasts, respectively. They give in-depth reviews with the state-of-the-art information. Chapter 3 deals with genetics of methylotrophs. It records the up-to-date progress and points out areas where future successes might lie.

The book then turns from the basic sciences to the appllied aspects of methylotrophs. Chapters 4, 5, and 6 were written by scientists from industries which have an active research program on methylotrophs.. Chatper 4 reviews production of secondary metabolites by methylotrophs, including amino acids, vitamins and coenzymes, and polysaccharides. Chapter 5 is devoted to methylotrophs in single cell production, the initial interest for industrial application. Other applied aspects of methylotrophs are summarized in Chapter 6. It includes production of specialty chemicals, pollution clean up, and cofactor regeneration. In addition, it includes a patent up-date covering some applied aspects of methylotrophs starting from 1960 to the present.

This book is intended for reference purposes at the professional level and for students at the graduate level. It is hoped that it will provide researchers with not only basic science, but also applied aspects of methylotrophs.

Ching T. Hou
Linden, N. J.

CONTRIBUTORS

Yoshio Hirose, Ph.D.
Ajinomoto Company, Inc.
Central Research Laboratories
Kawasaki, Japan

B. W. Holloway, Ph.D., D.Sc.
Professor of Genetics
Monash University
Clayton, Victoria
Australia

Ching T. Hou, Ph.D.
Research Associate
Exxon Research Center
Annandale, New Jersey

Yasushi Morinaga, Ph.D.
Ajinomoto Company, Inc.
Central Research Laboratories
Kawasaki, Japan

K. A. Powell, Ph.D.
Imperial Chemical Industries
Agricultural Division
Microbiology Section
Cleveland, United Kingdom

B. L. F. Rodgers, Ph.D.
Chemical Engineer
Imperial Chemical Industries
Agricultural Division
Cleveland, United Kingdom

Yoshiki Tani, Ph.D.
Professor
Research Center for Cell and Tissue Culture
Faculty of Agriculture
Kyoto University
Kyoto, Japan

TABLE OF CONTENTS

Chapter 1

MICROBIOLOGY AND BIOCHEMISTRY OF METHYLOTROPHIC BACTERIA

Ching T. Hou

TABLE OF CONTENTS

I. INTRODUCTION

Methylotrophic bacteria are recognized by their ability to use, as sole carbon and energy sources for growth, compounds that contain no carbon-carbon bonds and to assimilate carbon as formaldehyde or a mixture of formaldehyde and carbon dioxide.[1-6] They are different from methanogenic bacteria which produce methane from organic compounds. Growth substrates often used by methylotrophs are the following: methane, methanol, formaldehyde, formate, dimethyl ether, methyl formate, formamide, and methylamines. Not all of the methylotrophs can grow on methane. Those microorganisms capable of growth on methane are called methanotrophs. No obligate methylotrophs so far described are capable of growth on formate. Facultative methylotrophs can also grow on a variety of organic multicarbon compounds.

Methylotrophs play an important role in carbon recycling in the biosphere. The stability of the biosphere carbon cycle depends on the uninterrupted balance among methanogens which produce C_1-compounds from organic compounds anaerobically, methylotrophs which oxidize C_1-compounds to carbon dioxide or cell mass aerobically, and heterotrophs which degrade methylotroph biomass. Much of the methane arising from methanogenesis in the deeply submerged sediment does not reach the atmosphere due to its aerobic and anaerobic oxidation by methane-utilizing bacteria in both aquatic and soil environments.[7]

The progress in the understanding of the microbiology and biochemistry of methylotrophs can be attributed largely to Whittenbury and co-workers who isolated and characterized many methane-oxidizing bacteria,[8] and to Quayle and co-workers who unveiled the metabolic pathways of C_1 compounds in methylotrophs.[2] However, the importance of specific methylotrophic microorganisms in nature has not been established. We cannot be certain that the dominant methylotrophs in nature have been isolated and characterized.

The purpose of this chapter is to review in detail up-to-date information on microbiology and biochemistry of methylotrophic bacteria. The following excellent reviews were often used as sources of information during the preparation of this chapter: Quayle,[2,3] Anthony,[4] Colby et al.,[5] Hanson,[7] Wolf and Higgins,[9] and Higgins et al.[10]

II. MICROBIOLOGY

A. Taxonomy

The first methane-utilizing microorganisms, *Bacillus methanicus,* was isolated by Söhngen in 1906.[11] The name was changed to *Methanomonas methanica* by Orla-Jensen,[12] to *Pseudomonas methanica* by Dworkin and Foster,[13] and finally to *Methylomonas methanica* by Foster and Davis.[14] Aside from this *M. methanica,* only three additional species were well documented prior to 1970. They are *Pseudomonas methanitrificans,*[15] *Methylomonas methanooxidans,*[16,17] and *Methylococcus capsulatus.*[14] In 1970, Whittenbury and co-workers isolated over 100 methane-utilizing bacteria and classified these bacteria into several groups on the basis of morphology, fine structure, and type of resting stage formed. All of the strains isolated were Gram-negative, catalase-positive, aerobic bacteria that utilized compounds without carbon-carbon bonds as substrates. All were obligate methylotrophs utilizing methane, methanol, and dimethylether as carbon and energy sources.[8] Many of the isolates formed heat- or desiccation-resistant resting stages. All of the strains contained complex intracytoplasmic membrane structures and utilized different carbon assimilation pathways.[2,18-20] Based on the differences in intracytoplasmic membrane arrangement and in carbon assimilation pathways, Whittenbury et al. classified the isolates into two types (Table 1). Most workers accept this as a basis for the development of a formal classification, and most organisms isolated since then fall more or less into one of these groups. Romanovskaya et al.[21] have reviewed the information about genera and species of methylotrophs and have

Table 1
CLASSIFICATION OF METHANE-UTILIZING BACTERIA

Intracytoplasmic membrane type	Group	Morphology	Resting stage
I	Methylomonas	Rod	Immature azotobacter-type
I	Methylobacter	Rod	Azotobacter-type cyst
I	Methylococcus	Coccus	Immature azotobacter-type
II	Methylosinus	Rod or pear shaped	Exospore
II	Methylocystis	Rod or vibroid	Lipid cyst

corrected nomenclature where it did not follow the project for an International Code of Bacterial Nomenclature. As more detailed knowledge has accumulated, the need for two subgroups (A and B) above the genus level has arisen for classification of type I organisms. The new genus of facultative methylotrophs, *Methylobacterium*[22,23] needs to be accommodated. A tentative classification scheme for methane-oxidizing bacteria including obligate and facultative methylotrophs was proposed (Table 2). This scheme still divides the bacteria into the major groups (type I and type II) and subgroups. The validity of this classification scheme has not been adequately tested. Not all methane-oxidizing bacteria are known to possess all the characteristics of a proposed group or subgroup. Some type I methane-oxidizing bacteria that assimilate most of their carbon by the ribulose monophosphate pathway also contain low levels of enzymes of the serine pathway for formaldehyde fixation.[5,24] Hydroxypyruvate reductase has been found in extracts of *Methylococcus capsulatus* and *Pseudomonas methanica*.[5,25] Serine-glyoxalate aminotransferase was found in several type I methane oxidizers.[5,21,26] The metabolic function of these enzymes in type I methylotrophs is unknown. The presence of three pathways (ribulose monophosphate, ribulose diphosphate, and serine pathways) for the incorporation of C_1 units in *M. capsulatus* (Bath) is quite unusual.[27] However, Bamford and Quayle[28] have pointed out that the presence of one or two enzymes of a pathway in an organism does not indicate that the whole pathway functions. There is no published evidence for the simultaneous functioning of both pathways for the formaldehyde fixation in a methylotroph. The role of secondary alcohol dehydrogenase[29] and 1,2-propanediol dehydrogenase[288] in methylotrophs is also not clear.

Most type I and type II methane oxidizers are capable of fixing dinitrogen.[5,30-32] Despite all of these variations, the major division of methane-oxidizers into two groups based on the major pathways utilized for formaldehyde assimilation as well as other properties remains valid.

Methylotrophs also include bacteria and fungi that grow on methanol and methylamines but do not oxidize methane. These organisms also grow on heterotrophic substrates.[2,4,5] A comparison of assimilatory pathways utilized by some methylotrophs that also grow on heterotrophic substrates is listed in Table 3. There is a greater diversity in morphology, physiological capabilities, and one-carbon assimilatory pathways among these bacteria than among the methane utilizers. Whereas all methane utilizers are Gram-negative bacteria, methanol utilizers can be either Gram-positive or Gram-negative. The morphological diversity includes the stalked, budding bacteria of the genus *Hyphomicrobium, Pseudomonas, Streptomyces, Mycobacteria,* and spore-forming bacilli.

B. Morphology

Methanotrophic bacteria differ in morphology. There are rods, cocci, vibrioid and pear-shaped organisms of various sizes and dimensions (Table 1). Many form resting stages, of which there are three distinct types: exospores, and two types of cyst; one is lipid cyst and the other azotobacter-type cyst.[33] Exospores are formed by two varieties of methane-utilizer,

Table 2
TENTATIVE CLASSIFICATION SCHEME FOR METHANE-OXIDIZING BACTERIA[a]

Determinants	Type I		Type II
Membrane arrangement	Bundles of vesicular disks		Paired membranes around cell periphery
Resting stages	Cysts (azotobacter-like)		Exospores or lipid cysts
Major carbon assimilation pathway	RMP (3-hexulosephosphate synthase)		Serine pathway (hydroxypyruvate reductase + 3-hexulosephosphate synthase)
TCA cycle	Incomplete (2-oxoglutarate dehydrogenase negative)		Complete
Nitrogenase	Some +		+ (those tested)
Predominant fatty acid C chain length	16		18
	Subgroup A	**Subgroup B**	**Subgroups obligate[b] and facultative[c]**
Autotrophic CO_2 fixation	–	+	–
DNA base ratio (% G + C)	50—54	62.5	62.5+ (where tested)
Isocitrate dehydrogenase	NAD or NADP dependent	NAD dependent	NADP dependent
Cell shape	Rod & coccus	Coccus	Rod and vibrio
Growth at 45°C	Some +	+	–
Examples	*Methylomonas methanica* and *M. albus*	*Methylococcus capsulatus*	*Methanomonas methanooxidans. Methylosinus trichosporium* (both obligate) and *Methylobacterium organophilum* (facultative)

[a] Not all strains classifiable into type I and type II have been shown to possess all the biochemical characteristics outlined in this scheme.
[b] Use methanol and formaldehyde as carbon and energy source, but not C_{2+} compounds.
[c] Use variety of organic compounds, e.g., glucose as carbon and energy source.

From Colby, J., Dalton, H., and Whittenbury, R., *Ann. Rev. Microbiol.*, 33, 481, 1979. With permission.

Table 3
CARBON ASSIMILATION AND CATABOLIC PATHWAYS UTILIZED BY FACULTATIVE METHYLOTROPHS

Organism	Substrate used for growth	C_1-Assimilation pathway
Methane-oxidizing bacteria		
Methylobacterium organophilum XX	Methane, methanol	Icl^-, serine
Methylobacterium organophilum R6	Methane	Serine
Bacteria unable to grow on methane		
Pseudomonas AM1	Methanol, methylamines	Icl^- serine
Organism JB1	Methanol	Icl^- serine
Hyphomicrobium spp.	Methanol, methylamines	Icl^- serine
Organism JB1	Methanol	Icl^- serine
Bacterium 5B1	Methylamines	Icl^- serine
Pseudomonas MA	Methylamines	Icl^+ serine
Pseudomonas aminovorans	Methylamines	Icl^+ serine
Pseudomonas MS	Methylamines	Icl^+ serine
Bacillus spp. PM6		Ribulose monophosphate
Pseudomonas C	Methanol	Ribulose monophosphate
Pseudomonas oleovorans	Methanol, methylamines	Ribulose monophosphate
Arthrobacter globiformis	Methylamines	Ribulose monophosphate
Streptomyces spp.	Methanol	Ribulose monophosphate and serine
Brevibacterium sp. 24	Methylamines	Ribulose monophosphate
Mycobacterium sp. 10	Methylamines	Ribulose monophosphate

From Hanson, R. S., *Adv. Appl. Microbiol.*, 26, 3, 1980. With permission.

Methylosinus trichosporium and *M. sporium*. These exospores are finally budded off. Spores formed by all strains are desiccation-resistant, surviving at least 18 months in a dried state in the absence of methane, and heat-resistant, surviving 85°C for 15 min. Freshly formed spores are not refractile and can be stained with Gram and other conventional simple stains. On aging (1 to 7 days), many spores become acid-fast, refractile, and can only be stained with endospore stains.

Lipid cysts are formed by only one organism, *Methylocystis parvus*. The lipid inclusion formed as the organism enters the stationary phase is mainly poly-β-hydroxybutyrate. The lipid organism is not heat-resistant (cannot survive 65°C for 30 min) but is desiccation-resistant (viable after 1 week dried on a glass slide).

All varieties of methanotrophs form multiple-, double-, and single-bodied cysts, indistinguishable in microscopic appearance from cysts formed by *Azotobacter* species. No factor has been found which stimulates cyst formation, though a correlation has been observed between a low level of oxygen and an increase in encystment rate. Cyst formation in some strains is prevented by the omission of magnesium ions from the medium. In many cases, profuse encystment is accompanied by the production of a pigment ranging in color from yellow, to rust to brown, to dark-brown. Nonencysted cultures remain creamy white. These azotobacter-type cysts are not heat-resistant but are desiccation-resistant. Mature cysts can survive at least 3 months in a dried state in the absence of methane.

A third variety of the resting stage, an immature cyst, is also recognized. Immature azotobacter-type cysts are found in methane-utilizers which do not form desiccation-resistant cysts. They belong either to the *Methylomonas* (rods) or *Methylococcus* (cocci) group. These cysts survive the absence of methane for 5 to 6 weeks in contrast to vegetative organisms which survive only 2 to 3 days.

C. Fine Structure

All methanotrophic bacteria are found to possess complex internal membranous structures. Two basic membrane arrangements are recognized; type I is found in the rod-coccoid organisms (*Methylomonas*, *Methylococcus*, and *Methylobacter* groups) and type II is found in the vibrioid organisms (*Methylosinus* and *Methylocystis* groups).[18]

1. Type I

Type I methanotrophs possess vesicular discs of membrane organized into bundles which are distributed throughout the organism (Figure 1).[16,18,34-38] The individual membranes are about 90 Å thick, consist of two outer electron-dense layers and an inner electron-transparent layer, and are similar in size and appearance to the cytoplasmic membrane. The membranes are a constant feature of these bacteria when grown on either methane or methanol.[18,36,39]

The possibility that they have been induced by ammonia, a methane analog, could not be completely ruled out.[40] However, methanol-grown methylotrophs are largely filled with electron-lucent droplets. When they are transferred back to grow on methane, it induces membrane formation concomitant with a reduction in the number of electron-lucent droplets.[36] It has thus been concluded that intracytoplasmic membranes are induced by methane per se.

Hyder et al.[36] also demonstrated that these membranes are affected by culture age and conditions. Exponentially growing organisms exhibit the classical orderly parallel stacks (bundles) of membrane. As the culture enters into the early stationary phase, these membrane bundles become less numerous and less orderly. Many organisms contain electron-lucent droplets. In the late stationary phase, this disordered state becomes even more pronounced. The membrane sacs are widely distended, if present at all, and many spheroidal inclusions are found.

All detailed studies of type I methanotrophs have used organisms grown on shake flasks, i.e., under limited tension of methane and oxygen. DeBoer and Hazeu[34] have observed the fine structure of a methanotroph grown in a fermentor. They noted that under these conditions, the membrane stacks were less extensive. Monosov[40] suggested that membrane genesis is connected with the accumulation of phospholipid micelles in growing organisms. The centers of micelle formation are closely associated with the cytoplasmic membrane and may be precursors of the intracytoplasmic bilayers.

2. Type II

Type II methanotrophs possess paired membranes running throughout the bacterium or aggregated at its periphery where they run parallel to the cytoplasmic membrane. This type of membrane, appearing when the organisms are grown on methane or methanol, are more variable and less ordered than that of the rod-coccoid strains. The individual membranes are about 80 Å thick and similar to the cytoplasmic membrane in both size and appearance. The membranes are always in pairs and bound a lumen of varying dimensions which is usually filled with material noticeably less electron-dense than the surrounding cytoplasm. Branching of the membrane is sometimes observed. Two groups of methanotrophs possess type II membrane structure: *Methylosinus* and *Methylocystis* (Figure 2). Extensive peripheral flattened vesicles have been observed in *Methylosinus trichosporium* OB3b.[41]

M. trichosporium OB3b also shows marked changes in intracytoplasmic membrane content and internal morphology which are dependent on growth conditions.[42,44] During growth in shake flasks, the membrane content appears to be related to the amount of oxygen in the gas phase.[44] Organisms grown with little air or oxygen show less membrane. Growth in a continuous culture at 30°C under a variety of conditions, for example, methane, oxygen, or nitrogen limitation or at maximum growth rate, yields organisms lacking parallel paired membranes and containing only membrane-bound vesicles.

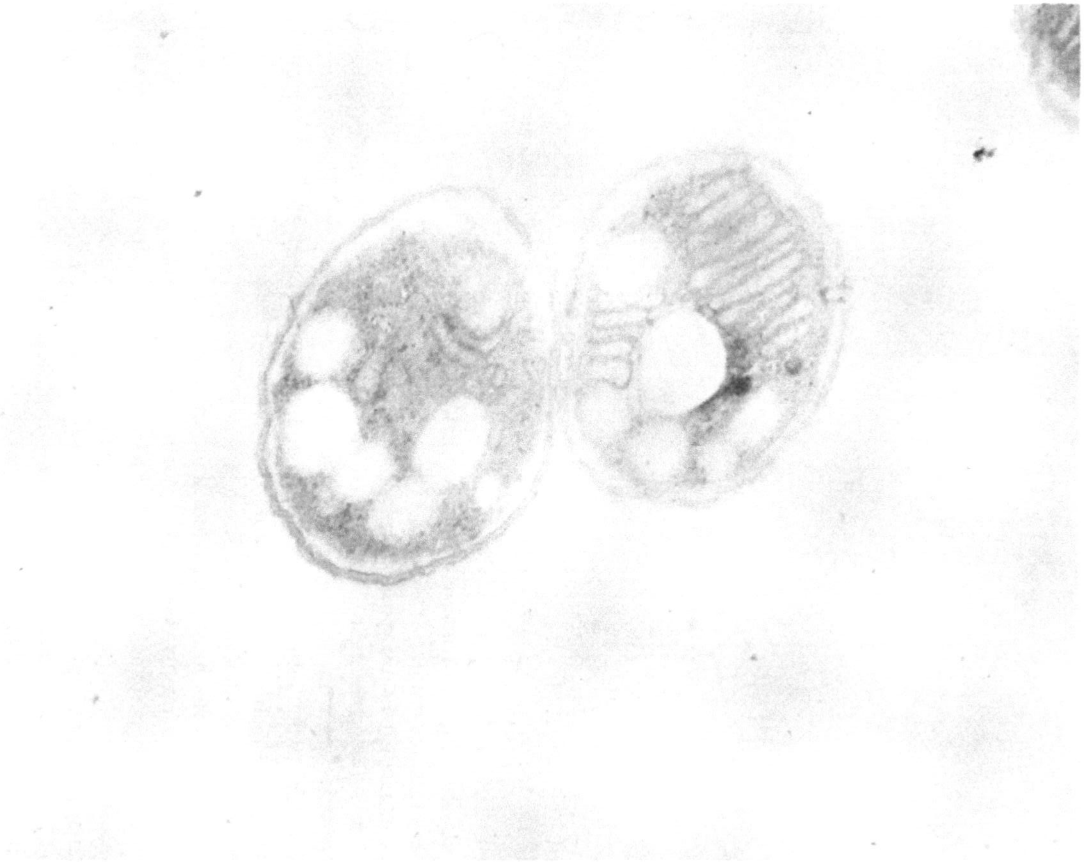

FIGURE 1. Thin section electron micrograph of *Methylococcus* sp. grown on methane.

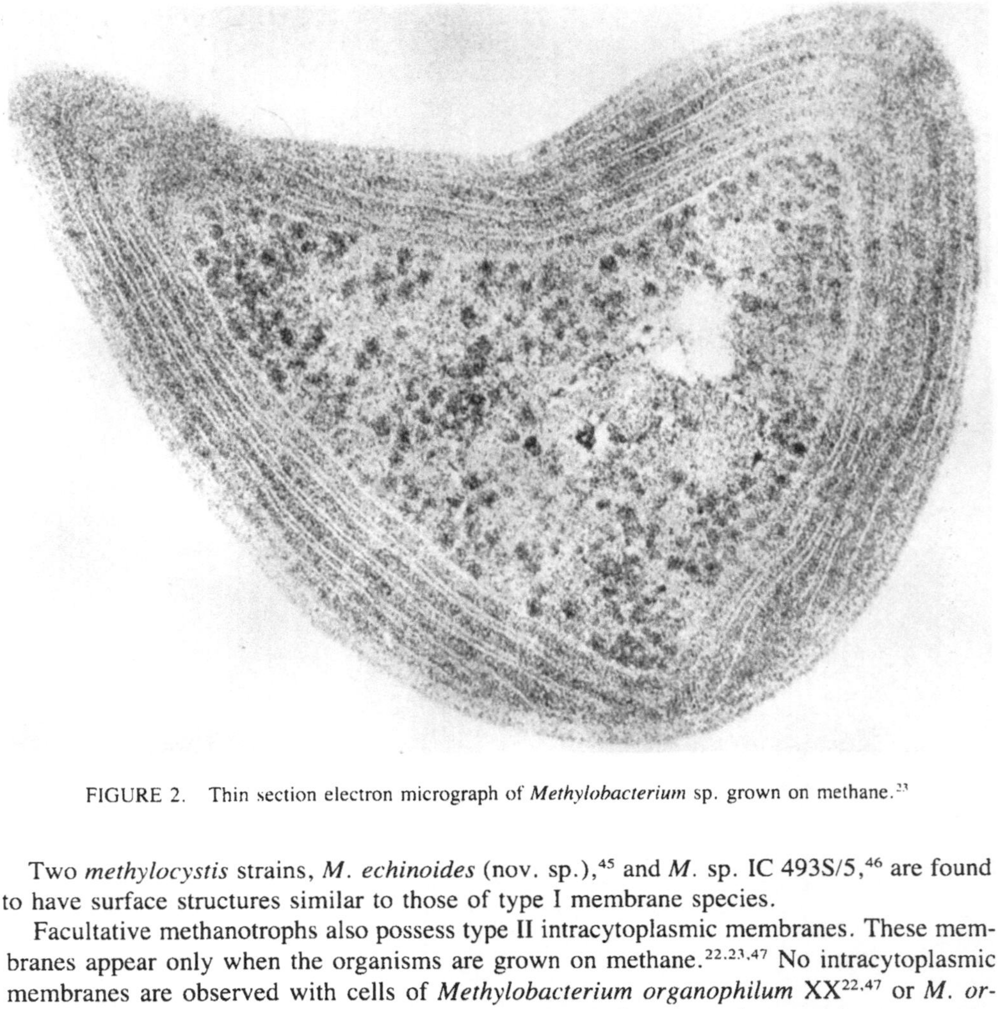

FIGURE 2. Thin section electron micrograph of *Methylobacterium* sp. grown on methane.[23]

Two *methylocystis* strains, *M. echinoides* (nov. sp.),[45] and *M.* sp. IC 493S/5,[46] are found to have surface structures similar to those of type I membrane species.

Facultative methanotrophs also possess type II intracytoplasmic membranes. These membranes appear only when the organisms are grown on methane.[22,23,47] No intracytoplasmic membranes are observed with cells of *Methylobacterium organophilum* XX[22,47] or *M. organophilum* CRL 26[23] when organisms are grown on methanol or other organic compounds. The repression/induction of intracytoplasmic membrane synthesis is correlated with the inducibility of methane oxidation. When strain XX is grown on methane as its sole carbon and energy source, the cells have intracytoplasmic membranes, and the resting cells will oxidize methane. By transferring strain XX to any other carbon substrate, the intracytoplasmic membrane synthesis for methane oxidation is repressed. When the cells are transferred back to the methane substrate, the intracytoplasmic membranes are synthesized, and the cells oxidize methane. Similar phenomena are observed in strain R6 (CRL 26).[23] These observations lead to a speculation that intracytoplasmic membranes are induced by methane and are involved in methane oxidation.

D. Role of Intracytoplasmic Membranes

The role of these extensive membrane systems in both type I and II organisms has not yet been clearly elucidated. Davies and Whittenbury[18] have ruled out the possibility that low solubility of the substrate and metabolic stress resulting from the synthesis of components from C_1-compounds are responsible for the development of such organelles. In several methylotrophs it appears that methane per se does not induce membrane formation.[18] Contrary to the reports on facultative methylotrophs[22,23,47] in which no intracytoplasmic membranes are observed when cells are grown on methanol or other organic compounds, methanol-grown organisms under certain conditions possess intracytoplasmic membranes.[18,39,42] It is possible that these membranes are synthesized in response to a methane analog, ammonia or methanol.[39,42] In the facultative methylotroph, *Methylobacterium organophilum* XX,[48] the content of intracytoplasmic membrane is dependent on both carbon source and dissolved oxygen tension. At very low dissolved oxygen tension, three to four layers of paired membrane are present, whereas at higher dissolved oxygen tension, only one peripheral paired membrane is observed. No membrane formation is observed during growth on methanol, glucose, or succinate at similar low dissolved oxygen tension. Synthesis of the membrane is also encouraged by low temperature and low growth rates. As has been mentioned previously in type II microorganisms, the intracytoplasmic membrane content of *Methylosinus trichosporium* OB3b also depends on the amount of oxygen in the gas phase.[44] The association of these membranes with methane monooxygenase activity has been speculated.[18] A number of the enzymes involved in methane oxidation, including methane monooxygenase,[49,53] methane oxidase and methanol dehydrogenase,[54] and formaldehyde oxidase and formaldehyde dehydrogenase, are particulate.[54] The requirement for phospholipase or detergents for the solubilization of methane hydroxylase suggests that methane monooxygenase is an integral membrane protein of *M. trichosporium* OB3b.[53] The membrane matrix could provide a favorable interface for the accommodation of methane and the methane monooxygenase because the hydrophobic core of the intracytoplasmic membrane will maintain a higher concentration of methane in the vicinity of the enzyme. Recent findings further confirm the relationship between intracytoplasmic membrane content and the nature of methane monooxygenase. Methane monooxygenase is located in the particulate fraction in organisms grown in a shake flask which has limited oxygen tension.[44,45,56] Whereas in fermentor-grown organisms which have high oxygen content and have less intracytoplasmic membranes, the methane monooxygenase is in soluble form.[42,56] It is now generally believed that the internal morphology of methanotrophs is dependent upon growth conditions.[36,42,44,57]

E. Classification

Methanotrophs have been divided into two major groups type I and type II,[18] based on their internal membrane structures. This major division is supported by the findings of Lawrence and Quayle[58] which state that methanotrophs of different membrane types employ different carbon incorporation pathways. The need for two subgroups (A and B) above the genus level for type I organisms and for the creation of a new genus for facultative methylotrophs has been proposed (Table 2). Subgroup A of type I methanotrophs includes those bacteria with a DNA base ratio of 50 to 54 mol % G + C, are rods or cocci, and do not fix CO_2 autotrophically. *Methylomonas methanica* and *M. albus* are examples of organisms in this subgroup.

Subgroup B of type I methanotrophs includes those bacteria that have a DNA base ratio of 62.5 mol % G + C, that are cocci, that grow at 45°C, and that contain phosphoribulokinase plus ribulose diphosphate carboxylase, key enzymes of Bensen-Calvin cycle for autotrophic CO_2 fixation in addition to enzymes of ribulose monophosphate pathway for carbon assimilation. Isocitrate dehydrogenase is NAD^+ dependent. *Methylococcus capsulatus* is the type culture of this subgroup.

Subgroup obligate and subgroup facultative are the two subgroups proposed for type II methanotrophs.

Originally, Whittenbury et al.[8] classified methanotrophs into five groups (genera) based on the variation in morphology, fine structure, and type of resting stage formed: two vibroid groups, Methylosinus and Methylocystis; and three rod/coccoid groups, Methylomonas, Methylobacter, and Methylococcus. This classification of methanotrophs is still accepted by most workers in this field as a basis for the development of a formal classification.

Table 4
PROPERTIES OF SUBGROUPS OF METHANE-UTILIZING BACTERIA

Group and subgroup	No. of strains	Growth at 37°	Growth at 45°	Growth on methanol (0.1% w/v)	Shortest division time (hr)	Mobility and flagellation	Capsule formed
Type I bacteria							
Methylomonas							
methanica	30	−	−	+	3.5	+P	+
albus	3	+	−	+	3	+P	−
streptobacterium	5	−	−	−	4	−	+
agile	4	+	−	+	3.5	+P	−
rubrum	7	−	−	+	4	+P	+
rosaceus	2	−	−	−	4	+P	+
Methylobacter							
chroococcum	9	−	−	−	5	−	+
bovis	5	+	−	−	4	−	−
capsulatus	4	−	−	−	4	+P	+
vinelandii	5	+	+	−	3	+P	+
Methylococcus							
capsulatus	3	+	+	−	3.5	−	+
minimus	5	−	−	−	3.5	−	+
Type II bacteria							
Methylosinus							
trichosporium	9	+	−	−	5	+PT	+[a]
sporium	12	+	−	−	5	+PT	+[a]
Methylocystis							
parvus	1	+	−	−	5	−	+[a]

Note: PT = Polar tufts of flagella; P = polar flagellum;

[a] Capsules under electron microscope consisted of short fibers radiating from cell wall. No structure was seen in capsules of other organisms.

From Whittenbury, R., Phillips, K. C., and Wilkinson, J. F., *J. Gen. Microbiol.*, 61, 205, 1970. With permission.

Properties of subgroups of methane-utilizing bacteria are listed in Table 4. For the details of classification, the reader should consult the work of Whittenbury et al.[8]

1. Group Methylomonas

All the organisms are rod-shaped and possess a type I membranes structure. Many are capsulated. Motile strains are polarly flagellated. Some strains form immature azotobacter-type cysts which are not desiccation-resistant but survive the absence of methane for 4 to 5 weeks. Vegetative cells survive only 3 to 4 days. All organisms utilize the ribulose monophosphate pathway for major carbon assimilation. They have an incomplete tricarboxylic acid cycle lacking α-ketoglutarate dehydrogenase. The G + C content of the DNA base ratio of these organisms is 50 to 54 mol % lower than organisms of other groups. They do not fix CO_2 autotrophically. Subgroup methanica is the most commonly encountered methane-utilizing bacterium.

2. Group Methylobacter

Most strains are rod-shaped at all stages in their growth cycle, many resembling Methylomonas strains morphologically. Some are remarkably similar to the large-cell-forming species of *Azotobacter*, changing from rod form to coccal and intermediate forms and back to rod form. All possess type I membrane structures and utilize the ribulose monophosphate pathway for carbon assimilation. They also lack α-ketoglutarate dehydrogenase and have an incomplete tricarboxylic acid cycle. All the organisms form slime and capsule, desiccation-resistant cysts, and produce pigment on cyst formation.

3. Group Methylococcus

These organisms are nonmotile cocci possessing capsules and a type I membrane structure. They form resting stages similar to immature azotobacter cysts. Major carbon assimilation of these organisms is ribulose monophosphate pathway. However, the coexistence with other carbon assimilation pathways was reported.[27]

4. Group Methylosinus

The organisms are generally rod-shaped, but occassionally of bizarre form, and possess a polar tuft of flagella. Rosette formation is frequently observed. The capsules formed by these organisms do not respond to the polysaccharide stain. At the onset of sporulation, the organisms assume a pear-shape or vibrio-shape. Exospores which are heat-resistant, are budded off the nonflagellated poles of the organisms. All strains possess a type II membrane structure and utilize the serine pathway for their carbon assimilation. All organisms have a complete tricarboxylic acid cycle.

5. Group Methylocystis

The organisms are similar in fine structure (type II membrane structure) and rosettes formation habit to Methylosinus strains. However, they are nonmotile and nonspore-forming. Cells are rod or vibrioid-shaped. They form a cyst which is desiccation- but not heat-resistant. These organisms utilize the serine pathway for carbon assimilation and have a complete tricarboxylic acid cycle.

6. Group Methylobacterium organophilum

Cells are rod-shaped, Gram-negative and are normally found singularly with some rosettes. Negative stains indicate polar flagellation. The organisms utilize methane, methanol, succinate, and other organic compounds. The colonies produced by these organisms are light brown to pink, circular, and convex with entire margins. Cells possess type II intracytoplasmic membrane structure when they are grown on methane. The serine pathway for formaldehyde incorporation is the pathway of C_1 metabolism. The DNA base composition is 66 mol % guanine plus cytosine. Organisms have complete enzymes of the tricarboxylic acid cycle. Reported strains are *Methylobacterium organophilum* XX,[47] *M. organophilum* CRL 26,[23] *M. ethanolicum* and *M. hypolimneticum.*[59]

Many facultative methylotrophs which utilize methanol, methylamine, and other heterotrophic substrates have been isolated (Table 3). They are more diverse morphologically and physiologically, and include both Gram-positive and Gram-negative microbes. There had been no attempt made in grouping these facultative methylotrophs until 1982. Green et al.[60] studied 150 pink-pigmented facultative methylotrophs, 28 other facultative methylotrophs, and compared them with 16 nonmethylotrophic marker strains of the genera *Pseudomonas, Alcaligenes, Mycoplana,* and *Microcyclus.* There were 4 major and 7 minor clusters containing 187 of the 194 strains that were defined above the 80% similarity level. All of the organisms studied were divided into two distinct parts: nonpink-pigmented facultative methylotrophs, and pink pigmented facultative methylotrophs.

The new methane-utilizing facultative methylotrophs, *Methylobacterium organophilum*, was also examined by Green et al.[60] Similarities between these strains and other pink-pigmented facultative methylotrophs have been noted.[7] The ability to utilize methane was regarded originally as sufficient to warrant the allocation of the strain to a new genus, the definition of which excluded methane nonutilizing pink-pigmented facultative methylotrophs.

However, this property is plasmid-borne in *M. organophilum* and is easily lost.[61] Green and Bousfield[60,286] suggested that despite their apparent inability to utilize methane, pink-pigmented facultative methylotrophs should be placed under genus *Methylobacterium*.

F. Phospholipid and Fatty Acid Compositions of Methanotrophs

Galachenko and Suzina[38] have suggested that the fatty acid composition of methane-oxidizing bacteria is probably correlated with the structure of the intracytoplasmic membrane structures. *Methylococcus capsulatus* (a type I methanotroph) exhibits a typical Gram-negative phospholipid pattern, with phosphatidylethanolamine as the major phospholipid.[62] It also consists of phosphatidylglycerol, cardiolipin, and phosphatidyl choline. No intermediate methylated derivatives of phosphatidylethanolamine are found. The esterified fatty acids of type I methanotrophs are predominantly $C_{16:0}$ and $C_{16:1}$.[19,37,38,62] The monounsaturated fatty acids from *M. capsulatus* consist of both cis and trans isomers, predominantly Δ^9, Δ^{10}, and Δ^{11} double-bond positional isomers.[62]

Phospholipid patterns in *Methylosinus trichosporium*, a type II methanotroph, vary with each researcher. Weaver et al.[63] demonstrated a phospholipid pattern similar to a Gram-positive organism in that phosphatidylglycerol is the major component. Phosphatidylethanolamine and phosphatidylserine, were also observed. However, Makula[62] found that phosphatidylethanolamine was absent in *Methylosinus trichosporium*, and that the major phospholipid in this strain was methylated derivatives of phosphatidylethanolamine. These conflicts in phospholipid patterns have been attributed to the differences in cultivation methods. A striking feature of the phospholipid pattern between type I and type II methanotrophs is the similarity of the ratio between the acidic phospholipids and phosphatidylethanolamine; the ratio acids/PE + derivatives ranges from 1:4 to 5. The same ratio was observed in both whole cells and membrane preparations, indicating that the intracytoplasmic membrane composition is not unusual.[63] The major esterified fatty acid in type II methanotrophs is 18:1 (84 to 90%). The monounsaturated fatty acids from type II methanotrophs consist of both cis and trans isomers, predominantly the Δ^{10} double-bond positional isomer. The predominance of 16-carbon fatty acids in type I organisms and $C_{18:1}$ in type II organisms was first reported by Ribbons and co-workers.[37,63]

The presence of phosphatidylcholine, squalene, and sterols in bacteria with extensive intracytoplasmic membranes has been reported.[64,65] The phospholipid composition of a facultative methylotroph, *Methylobacterium organophilum*, was analyzed in order to determine whether there were any major difference in the cell grown under different conditions.[48] No qualitative difference was observed. Phosphatidylcholine was present as a major component of the phospholipids. Growth of the cells on methane results in a decrease in the ratio of phosphatidylcholine to phosphatidylethanolamine. Several sterols, differing from those identified in *Methylococcus capsulatus*,[64] were found in *Methylobacterium organophilum* grown on methane, methanol, or glucose. The greatest amount was found in cells grown on methane with a very low dissolved oxygen concentration, conditions favoring internal membrane formation. Sterols are believed to play an important structural role in membrane formation. However, in this study the sterols were not confined to the intracytoplasmic membrane material.

G. Tricarboxylic Acid Cycle

A key difference between type I and type II methane oxidizers is the presence of an incomplete or complete tricarboxylic acid cycle. Type I methanotrophs possess no detectable 2-ketoglutarate dehydrogenase.[66] Radioactive studies using ^{14}C-labeled acetate and the subsequent distribution of the label in amino acids of type I and type II organisms confirm the absence of 2-ketoglutarate dehydrogenase in type I strains and its presence in type II strains.[54,67] This difference continues to hold among new isolates of type I and type II obligate methanotrophs. It has been suggested that one basis for obligateness of methylotrophs is the lack of a complete TCA cycle, the absence of α-ketoglutarate dehydrogenase. This is unlikely to be true because type II obligate methylotrophs possess all the enzymes of the TCA cycle. The biochemical basis of obligateness of methylotrophs remains to be challenged.

III. BIOCHEMISTRY

A. Carbon Assimilation Pathways

Methylotrophs assimilate their carbon in the form of formaldehyde through either the ribulose monophosphate (RMP) pathway or the serine pathway. In general, type I methanotrophs utilize the RMP pathway whereas type II methanotrophs utilize the serine pathway. The existence of more than one carbon assimilation pathway in an organism has been reported. *Methylococcus capsulatus* has been shown to contain low levels of hydroxypyruvate reductase activity[25] and phosphoribulokinase and ribulose 1,5-bisphosphate carboxylase activities.[27] The presence of these key autotrophic carbon dioxide-fixing enzymes in this organism is quite interesting. Some type I organisms have also been shown to contain serine-glyoxylate transaminase activity[22] or hydroxypyruvate reductase activity.[68-71] Other serine pathway enzymes are also present in methane-grown *Methylococcus capsulatus* (Texas) and *Methylomonas methanica*.[72] In addition to enzymes of the autotrophic ribulose diphosphate pathway, *Paracoccus denitrificans* and *Thiobacillus novellus* also contain hydroxypyruvate reductase activity.[28]

From these observations, Bamforth and Quayle[29] concluded that the presence of merely one or two enzymes of a pathway in an organism cannot be used as evidence to conclude that such a pathway is operational. On the other hand, it is also unreasonable to reject the existence of another pathway merely because one major pathway is present. It is possible that carbon might be assimilated through more than one pathway in a methylotroph depending on environmental conditions.

The relationship between methylotrophs and autotrophs has been debated from two standpoints. One, from a more microbiological standpoint, is exemplified by Whittenbury and Kelly.[73] They compared and contrasted methylotrophs and autotrophs and concluded that the concept of autotroph might be extended to include all organisms able to assimilate C_1 compounds as their sole carbon source. The other, from a biochemical standpoint, is exemplified by Quayle and Ferenci.[74] They fear that the approach of Whittenbury and Kelly[73] could lead to an over-simplification of the issues. They are more cautionary and conservative; they emphasize the important biochemical pathways. Further evidence is required to resolve this issue.

1. The Ribulose Monophosphate Pathway

By using ^{14}C-labeled carbon dioxide, Leadbetter and Foster[75] reported that *Pseudomonas methanica* does not use the Calvin cycle for growth on methane. Johnson and Quayle[76] also showed no detectable ribulose diphosphate carboxylase activity in cell-free extracts of this bacterium. In subsequent studies with *P. methanica* and *Methylococcus capsulatus*[68,77] using ^{14}C-labeled methane, methanol, and formaldehyde, Quayle and co-workers showed that radioactivity was incorporated mainly into sugar phosphates, particularly hexose phosphate. It was found that cell-free extracts of both *P. methanica* and *M. capsulatus* catalyzed the condensation of formaldehyde with ribose 5-phosphate to form a mixture of sugar phosphates, the major one being fructose-6-phosphate. Detailed analyses of these works led to the proposal of a cyclic scheme for formaldehyde fixation analogous to the Calvin cycle.[2,77-79] Before Quayle suggested the name, ribulose monophosphate cycle, for this new formaldehyde fixation pathway, there were many misleading names used: the allulose phosphate pathway, the hexose phosphate pathway, the pentose phosphate cycle, and the ribose phosphate cycle.

The net synthesis of RMP pathway is one molecule of triose phosphate from three molecules of formaldehyde

$$3HCHO + ATP \rightarrow \text{triose phosphate} + ADP$$

2. The Enzymes of the Ribulose Monophosphate Pathway

The two key enzymes for the RMP pathway are 3-hexulose phosphate synthase and phospho-3-hexulose isomerase. Kemp[80] purified the membrane-bound enzyme 3-hexulose phosphate synthase from *M. capsulatus* and identified its substrates and products. It is known that this enzyme condenses formaldehyde and ribulose-5-phosphate to give D-arabino-3-hexulose phosphate. The second key enzyme in the pathway is phospho-3-hexulose isomerase which converts D-arabinose-3-hexulose-6-phosphate to fructose-6-phosphate. The second key enzyme in the pathway is phospho-3-hexulose isomerase which converts D-arabino-3-hexulose-6-phosphate to fructose-6-phosphate.

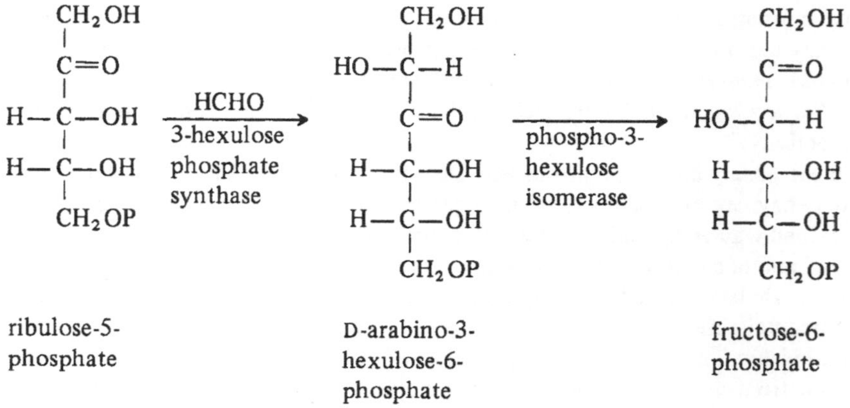

Both the synthase and the isomerase have been purified extensively from *Methylococcus capsulatus*,[81] and have molecular weights of 310,000 and 67,000, respectively. The synthase has subunits of molecular weight 49,000. The enzyme has also been purified from *Methylomonas* M15[82] and *M. aminofaciens* 77a.[83] Both the synthase and the isomerase have high substrate specificity.

The RMP pathway is conveniently divided into three stages (Figure 3). The details of the cycle differ somewhat between species, and the three variants are also shown in Figure 3.[5,77] Stage 1 involves the condensation of three molecules of formaldehyde with three molecules of ribulose-5-phosphate to yield three molecules of fructose-6-phosphate. Stage 2 involves the splitting of one molecule of fructose-6-phosphate to produce two molecules of C_3 compounds. In some methylotrophs, such as *Bacillus* spp. PM6 and S2A1[84] and *Arthrobacter globiformis* B-175,[85] this stage is achieved by enzymes of the glycolytic sequence.[77] In other methylotrophs, such as bacteria 4B6, C2A1, and W3A1,[84] *Pseudomonas* W6,[86] and *P. oleovorans*,[85] this split is catalyzed by enzymes of the Entner-Doudoroff variant and the analogous enzymes of the glycolytic sequence are absent. Some methanotrophs contain enzymes of both sequences and their relative importance of the two sequences in vivo is not known.[71] Stage 3 involves the regeneration of three molecules of ribulose-5-phosphate from the two molecules of fructose-6-phosphate and one molecule of glyceraldehyde-3-phosphate produced in stage 1 and 2. Transaldolase and transketolase catalyze the regeneration.[71,84] Some bacteria lack transaldolase and their regeneration of ribulose-5-phosphate involves sedoheptulose-1,7-diphosphatase and fructose diphosphate aldolase.[84]

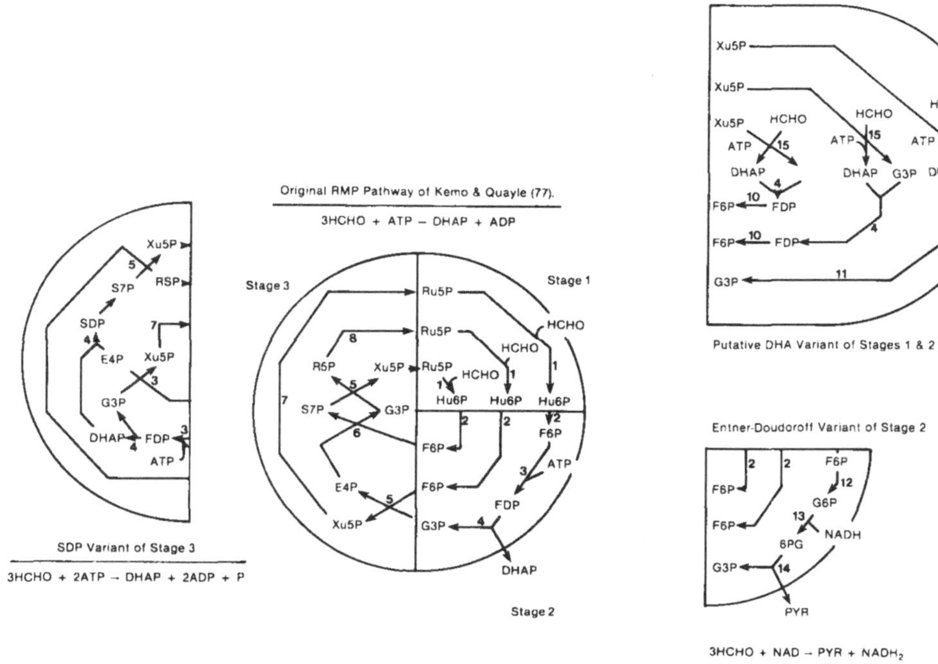

FIGURE 3. RMP and its variations. Ru5P, Ribulose-5-phosphate, Hu6P, D-erythro-L-glycero-3-hexulose-6-phosphate; F6P, fructose-6-phosphate, FDP, fructose-1,6-diphosphate; G3P, glyceraldehyde-3-phosphate; DHAP, dihydroxyacetone phosphate; E4P, erythrose-4-phosphate; Xu5P, xylulose-5-phosphate; S7P, sedoheptulose-7-phosphate; SDP, sedoheptulose-1,7-diphosphate; R5P, ribulose-5-phosphate; G6P, glucose-6-phosphate; 6PG, 6-phosphogluconate; pyr, pyruvate; 1, 3-hexulosephosphate synthase; 2, phospho-3-hexuloisomerase; 3, 6-phosphofructokinase (EC 2.7.1.11); 4, fructose diphosphate aldolase (EC 4.1.2.13); 5, transketolase (EC 2.2.1.1); 6, transaldolase (EC 2.2.1.2); 7, ribulose phosphate epimerase (EC 5.1.3.1); 8, ribulose phosphate isomerase (EC 5.3.1.6); 9, sedoheptulose diphosphatase; 10, fructose diphosphatase (EC 3.1.3.11); 11, triosephosphate isomerase (EC 5.3.1.1); 12, glucose phosphate isomerase (EC 5.3.1.9); 13, glucose-6-phosphate dehydrogenase (EC 1.1.1.49); 14, 6-phosphogluconate dehydratase (EC 4.2.1.12) + phospho-2-keto-3-deoxygluconate aldolase (EC 4.1.2.14); 15, transketolase + triokinase (EC 2.7.1.30). (From Colby, J., Dalton, H., and Whittenbury, R., *Ann. Rev. Microbiol.*, 33, 481, 1979. With permission.)

The presence of the two key enzymes of the RMP pathway, 3-hexulose phosphate synthase and phospho-3-hexulose isomerase, together with other evidence for the pathway have been found not only in methanotrophs but also in obligate methylotrophs and facultative methylotrophs (Table 5). Therefore, it is not specific either to methanotrophs or to obligate or facultative methylotrophs.

Strom, Ferenci, and Quayle[71] have pointed out that for those methylotrophs lacking dehydrogenases for formaldehyde or formate but which do have both 3-hexulose phosphate synthase and phospho-3-hexulose isomerase, the formaldehyde may be oxidized completely to carbon dioxide by a cycle involving glucose phosphate dehydrogenase and 6-phosphogluconate dehydrogenase (Figure 4). This cyclic oxidation of formaldehyde would generate NAD(P)H. Isotopic labeling studies indicate that not only does this cycle exist in vivo but it is the predominant mechanism for formaldehyde oxidation in the facultative methylotroph *Pseudomonas* C.[87] Recently, Beardsmore et al.[88] purified many enzymes from methanol-grown *Methylophilus methylotrophus*, which involves in this variant of the ribulose monophosphate cycle of formaldehyde fixation. The organism possesses high concentrations of a glucose-6-phosphate dehydrogenase active with both $NADP^+$ and NAD^+, and two separate 6-phosphogluconate dehydrogenases, one active with both $NADP^+$ and NAD^+ and the other

Table 5
METHYLOTROPHS USING THE RIBULOSE MONOPHOSPHATE PATHWAY

Methylotrophs	Ref.
Methane-oxidizing obligate	
Methylococcus capsulatus (Texas)	68, 71
M. capsulatus (Bath)	25, 137
M. minimus	53, 115
M. ucrainicus[a]	70
M. thermophilus[a]	70
Methylomonas methanica[a]	75, 71
M. agile	53
M. albus	115
M. mobilis	35
M. rosaceus	53
M. rubrum[a]	70
M. sp.	34
Methylobacter capsulatus	53
M. bovis[a]	70
M. vinelandii[a]	
Other obligate	
Bacterium 4B6	6, 84
Bacterium C2A1	84
Pseudomonas W6	86
Methylomonas M15	82
Methylomonas aminofaciens 77a	83
Facultative (not methane oxidizing)	
Bacillus spp. PM6 & S2A1	84
Arthrobacter 2B2	84
Arthrobacter globiformis B-175	247
Brevibacterium sp.	25
Mycobacterium sp.	14
Pseudomonas C	147
Pseudomonas oleovorans	246
Streptomyces sp. 239	69

[a] One or more enzymes of serine pathway also present.

From Colby, J., Dalton, H., and Whittenbury, R., *Ann. Rev. Microbiol.*, 33, 481, 1979. With permission.

active only with NAD^+. In addition, the organism contains methanol dehydrogenase, and NAD^+-linked formaldehyde and formate dehydrogenases; thus possesing the enzymic potential necessary for both the cyclic and linear sequences for the oxidation of the formaldehyde derived from methanol.

Ben-Bassat et al.[89] examined the distribution of methanol carbon between assimilation and oxidation pathways in methanol-grown *Pseudomonas* C, a facultative methylotrophic bacterium. In this strain, methanol is assimilated via the 2-keto-3-deoxy-6-phosphogluconate variant of the ribulose monophosphate pathway of formaldehyde fixation (Figure 4). There are three pathways of methanol metabolism to CO_2 in *Pseudomonas* C: the direct oxidation pathway which involves a formate as an intermediate; the cyclic oxidation pathway (glucose monophosphate shunt); and other decarboxylation reactions. It was concluded, by using radioactive compounds, that 25% of the methanol consumed by the cells was oxidized

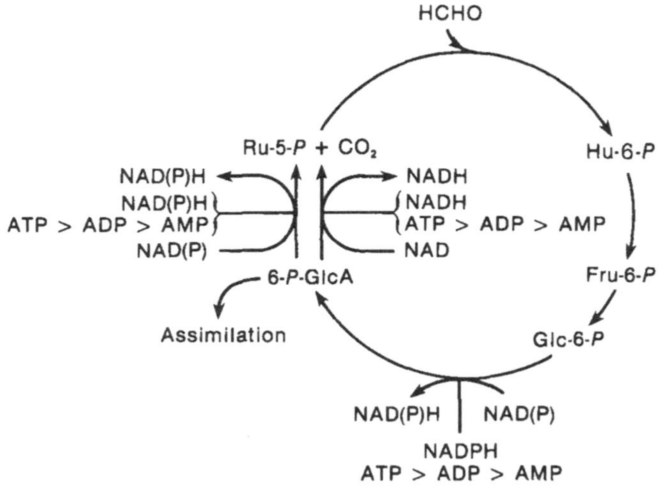

FIGURE 4. Scheme for cyclic oxidation of formaldehyde in *M. methylotrophus* (X-signifies inhibition). (From Beardsmore, A. J., Aperghis, P. N. G., and Quayle, J. R., *J. Gen. Microbiol.*, 128, 1423, 1982. With permission.)

through formate to CO_2 while the remainder was diverted into the hexulosephosphate synthase reaction from which 55% was assimilated through 2-keto-3-deoxy-6-phosphogluconate pathway, and 17% was oxidized to CO_2 via a cyclic oxidation pathway and other decarboxylation reactions. The remaining 7% from the methanol carbon was reincorporated as CO_2 into cell material through carboxylation reactions.

For some years it was thought that the classification of methylotrophs as type I or type II based on morphological criteria corresponded to their modes of carbon assimilation.[2] In all the methylotrophs examined, the dual presence of key enzymes of both assimilation pathways could not be demonstrated.[58] Recently, however, sufficient evidence has accumulated to show that some of the methylotrophs which utilize one pathway for carbon assimilation also contain enzymes for other pathways. For example some of the methylotrophs that use the RMP pathway for their carbon assimilation contain hydroxypyruvate reductase, an enzyme normally associated with the serine pathway.[58,69-71] This leads to one possibility that some organisms may use more than one pathway for carbon assimilation either under different growth conditions or even simultaneously. However, simultaneous operation of two major carbon assimilation pathways in any microorganism has not been demonstrated. Some reports demonstrated the presence of an auxiliary pathway. For example, methane-grown *Methylococcus capsulatus* (Bath), which uses the RMP pathway as the major route for carbon assimilation, also uses a serine pathway for recycling phosphoglycolate formed by the reaction of ribulose biphosphate carboxylase in the organism.[27] Trotsenko[90] examined 12 type I methylotrophs and found that only one of them solely utilized the RMP pathway. The rest of them also possess hydroxypyruvate reductase and serine-glyoxylate aminotransferase.

3. The Serine Pathway

The serine pathway was first proposed in 1961 by Large, Peel, and Quayle as a result of their work with methanol-grown facultative methylotrophs, *Pseudomonas* AM1 and *Hyphomicrobium vulgare*.[91,92] Short-term incubations of these facultative methylotrophs with ^{14}C-labeled methanol and carbonate resulted in early labeling of serine followed by glycine,

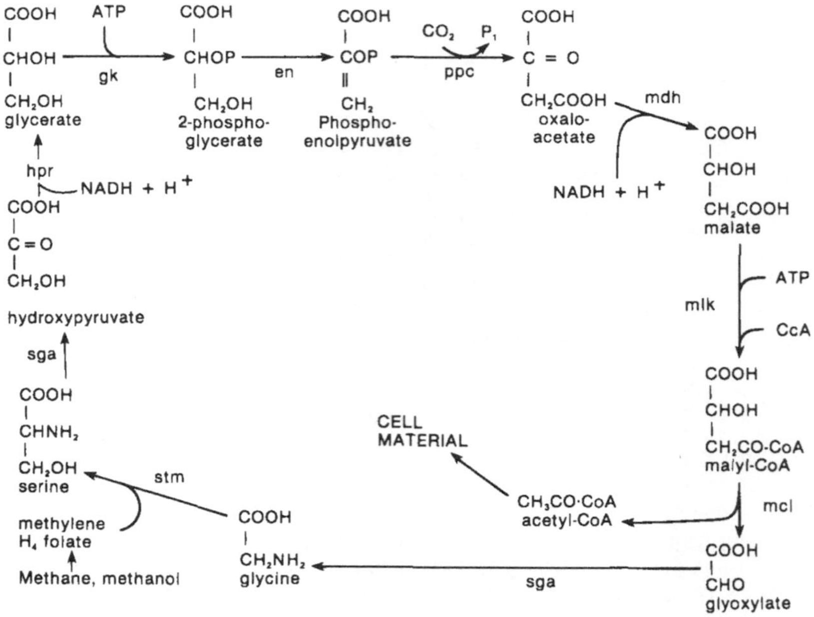

FIGURE 5. Reactions of the serine pathway: en, enolase; gk, glycerate kinase; hpr, hydroxypyruvate reductase; mel, malyl-CoA lyase; mdh, malate dehydrogenase; mtk, malate thiokinase; PPC, phosphoenolpyruvate carboxylase; sga, serine glyoxylate aminotransferase; stm, serine transhydroxymethylase. (From Wolfe, R. S. and Higgins, I. J., *Int. Rev. Biochem.*, 21, 267, 1979. With permission.)

malate, and aspartate. Two routes for formaldehyde assimilation were proposed in which the main difference was the synthesis of glycine: direct synthesis of glycine vs. cyclic regeneration of glycine. In both routes, half of the cell carbon arises by way of CO_2 produced by the complete oxidation of methanol. The key enzyme in both cases is serine transhydroxymethylase which catalyzes the incorporation of formaldehyde (as methylenetetrahydroforate). 5,10-Methylenetetrahydrofolate + glycine $\leftrightharpoons$ serine + tetrahydrofolate. Glycine acts as the formaldehyde acceptor. Serine transhydroxymethylase has also been detected in facultative methylotrophs,[2] in obligate methanotroph *Methylosinus trichosporium*[93] and in facultative methanotroph *Methylobacterium organophilum*.[94] Two transhydroxymethylase activities, one predominant during growth on methanol and other predominant during growth on succinate, were purified from the facultative methanotroph *M. organophilum* XX.[94] These two isoenzymes differed in their physical properties, i.e., molecular weight, electrophoretic and sedimentation patterns. Both isoenzymes were competitively inhibited by glycine, but only the methanol-induced enzyme was stimulated by glyoxylate. Many reactions between serine and oxaloacetate were subsequently identified with cell-free extracts of *Pseudomonas* AM1.[95,96] Evidence for cyclic regeneration of glyoxylate and glycine from a four-carbon compound came from ^{14}C-labeled succinate experiments using whole cells of *Pseudomonas* AM1. Salem et al.[97] have shown that (1) exposure of cells growing on methanol to ^{14}C-succinate resulted in early labeling of glycine and then serine and (2) when 2,3-^{14}C-succinate was administered to cells growing on methanol, the serine was labeled predominantly in the C_2 atom, which is consistent with its being formed by hydroxymethylation of glycine. These experiments confirmed a cyclic serine pathway for the formation of a C_2 compound from formaldehyde and CO_2. The serine pathway is shown in Figure 5.

4. The Enzymes of the Serine Pathway

Serine-glyoxylate aminotransferase catalyzes the following reaction:

$$\text{serine} + \text{glyoxylate} \rightarrow \text{hydroxypyruvate} + \text{glycine}$$

This involves two conversions in the serine pathway. The first conversion is the formation of glycine from its precursor, glyoxylate; the second one is the conversion of serine to hydroxypyruvate. This enzyme is inducible by growth on methanol. It was first detected in *Pseudomonas* AM1.[2]

The next enzyme in the serine pathway is hydroxypyruvate reductase which catalyzes the following reaction:

$$\text{Hydroxypyruvate} + \text{NADH}_2 \rightarrow \text{Glycerate} + \text{NAD}$$

High activity of this inducible enzyme has been detected in extracts of *Pseudomonas* AM1,[2] in type II methanotrophs,[58] in facultative methanotroph *Methylobacterium organophilum,*[22] and in facultative methylotrophs.[91,92] It also catalyzes the reduction of glyoxylate to glycolate at a much slower rate. Glycerate is then converted to 2-phosphoglycerate by glycerate kinase (EC 2.7.1.31):

$$\text{Glycerate} + \text{ATP} \rightarrow \text{2-phosphoglycerate} + \text{ADP}$$

The enzyme has been demonstrated in methanol-grown *Pseudomonas* AM1[98,99] and *Hyphomicrobium* X.[99] Half of the 2-phosphoglycerate generated by formaldehyde fixation is converted by another enzyme, phosphoglyceromutase to the net product of the pathway, 3-phosphoglycerate. The remaining half of the 2-phosphoglycerate is converted by enolase to phosphoenolpyruvate.

Phosphoenolpyruvate is the acceptor molecule for the second carbon fixation step of the serine pathway. First evidence for the formation of malate by carboxylation of a C_3 compound derived from serine also came from ^{14}C-labeled methanol and bicarbonate studies of Large et al.[71] The enzyme has later been identified as phosphoenolpyruvate carboxylase[100,101] (EC 4.1. 1.31) which catalyzes this carbon dioxide fixation reaction:

$$\text{phosphoenolpyruvate} + CO_2 + H_2O \rightarrow \text{oxaloacetate} + \text{Pi}$$

Carbon dioxide is derived mainly from the oxidation of reduced C_1 substrates. It has been calculated that approximately 50% of the carbon derived from methanol is assimilated as carbon dioxide by the facultative methylotrophs *Pseudomonas* AM1.[91] Strom et al. predicted a higher carbon dioxide requirement for serine pathway microorganisms than for RMP pathway microorganisms.[71] Their view has been reinforced by the findings of Trotsenko[90] and Loginova and Trotsenko.[101] They found that phosphoenolypyruvate carboxylase activity in the former is much higher than those in the latter microorganisms. Phosphoenolpyruvate carboxylase activity was also found in a type II methanotroph, strain M102.[102] O'Connor[61] reported two isoenzyme of phosphoenolpyruvate carboxylase in *Methylobacterium organophilum*. These two isoenzymes were precipitated at different ammonium sulfate concentrations. The isoenzyme of methanol-grown cells is insensitive to acetyl-CoA, but the activity of the other one from succinate-grown cells can be stimulated by acetyl-CoA.

Malate dehydrogenase (EC 1.1.1.37) catalyzes the conversion of oxaloacetate to malate:

$$\text{oxaloacetate} + \text{NADH}_2 \rightarrow \text{malate} + \text{NAD}$$

It has been reported in both obligate and facultative methanotrophs[22,23] and in all methylotrophs which use the serine pathway.

Malate thiokinase (EC 6.2.1.15) activity has been detected in *Hyphomicrobium* X,[72] *Pseudomonas* MA,[72,103] *Pseudomonas* MS,[72] and bacterium 5H2.[104] It catalyzes the following reaction:

$$\text{Malate} + \text{ATP} + \text{CoA} \rightarrow \text{Malyl-CoA} + \text{ADP} + \text{Pi}$$

Initially, the malate cleavage enzyme was called malate: ATP lyase. The first C_4 cleavage reaction was demonstrated in methylamine-grown *Pseudomonas* MA. [103] Hersh and Bellion[103] observed the reversible ATP- and CoA-dependent cleavage of malate into glyoxylate and acetyl-CoA. Subsequently, this reaction was found to consist of two reactions catalyzed by malate thiokinase and malyl-CoA lyase.[105]

Malyl-CoA lyase (EC 4.1.3.24) which catalyzes the following reaction:

$$\text{Malyl-CoA} \rightarrow \text{glyoxylate} + \text{acetyl-CoA}$$

has been described by Hersh,[105] Salem et al.,[72] and Cox and Zatman.[104] This enzyme has been detected in a variety of methylotrophs, including not only type II methanotrophs but also type I methanotrophs. The role of this enzyme in RMP pathway organisms is not clear.

Further evidence for the operation of the serine pathway comes from the experiments on mutants lacking hydroxypyruvate reductase,[2] serine-glyoxylate aminotransferase,[2] glycerate kinase,[106] or malyl-CoA lyase.[4] These mutant strains do not grow methylotrophically. These four enzymes are thought to be regulated coordinately, and their syntheses are repressed by succinate or its metabolites.

Originally, the acetyl-CoA produced from the cleavage of a C_4-compound was thought to be used in biosynthesis of cell material. Like the typical heterotrophic bacteria, many facultative methylotrophs grown on acetate produce glyoxylate from acetyl-CoA. Isocitrate lyase activity has been found in *Hyphomicrobium* X,[99] *Pseudomonas* MA,[107] *Pseudomonas* MS and *P. aminovorans*,[108], and bacterium 5H2,[104] during methylotrophic growth. The isocitrate lyase-serine pathway for assimilation of methanol in *Hyphomicrobium* has been proposed by Harder et al.[99] in which the intermediate withdrawn for biosynthesis is 3-phosphoglycerte. Quayle[109] described this isocitrate lyase-serine pathway in three phases. The first phase involves the net synthesis of acetyl-CoA from a reduced C_1-unit and carbon dioxide (Figure 6). In phase 2, the acetyl-CoA is oxidized to glyoxylate by a series of enzymes: citrate synthase, aconitase, and isocitrate lyase. In phase 3, the resulting glyoxylate is converted to glycine for the synthesis of phosphoglycerate via the serine pathway.

Isocitrate lyase is not present in all facultative methylotrophs. For example, *Pseudomonas* AM1 has no detectable activity of this enzyme even when grown on substrates usually requiring operation of the glyoxylate cycle for their assimilation. Beside the isocitrate lyase route, there are two other routes being proposed for the generation of a second glyoxylate acceptor molecule from acetyl-CoA:

1. In facultative methylotroph *Pseudomonas* AM1, acetyl-CoA is oxidized directly to glyoxylate with glycolate as a possible intermediate (Figure 4).[4] Bellion[110] could not demonstrate any key enzyme activity of any other known biochemical routes for the oxidation of acetyl-CoA to glyoxylate, such as the glycerate (tartronic semialdehyde) pathway or the β-hydroxyaspartate pathway in *Pseudomonas* strain PAR, a nonisocitrate lyase serine pathway organism.

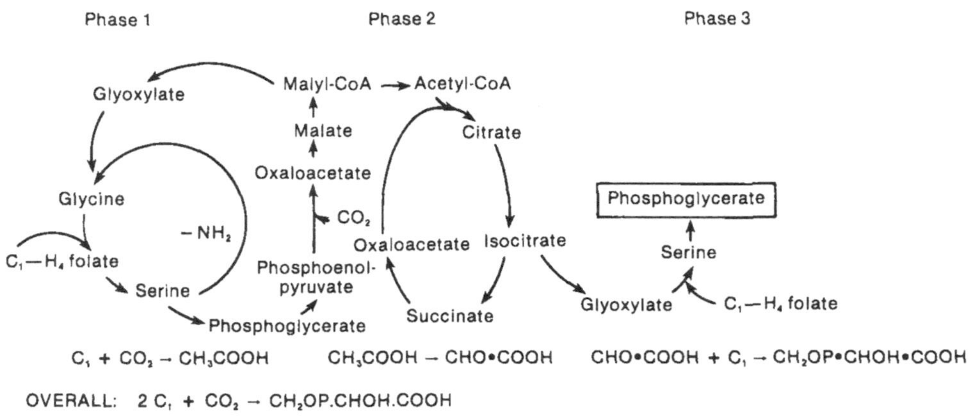

FIGURE 6. The isocitrate lyase-serine pathway. (From Quayle, J. R., *Microbial Growth on C_1 Compounds*, Terui, G., Ed., Society of Fermentation Technology, Tokyo, 1975, 59. With permission.)

2. Kortstee[111] proposed the homoisocitrate pathway in which acetyl-CoA is condensed with 2-oxoglutarate to give homocitrate. This was subsequently rearranged to give homoisocitrate. Cleavage of this C_7 molecule generated back to 2-oxoglutarate (Figure 5). This pathway was demonstrated only in facultative methylotroph *Pseudomonas* strain 80.

5. Production of Intracellular and Extracellular Polymers

Accumulation of poly-β-hydroxybutyrate (PHB) in *P. methanica* in large quantities (25% of the dry weight of 4-day-old cultures) was first observed by Kallio and Harrington.[112,113] Many of the methane-utilizing strains identified by Whittenbury et al. were packed with lipid inclusions, mainly PHB.[8,33] PHB was also found in vibrio shaped methane oxidizing bacteria[114] and in thermophilic and thermo-tolerant methane utilizers.[115] Methane-grown type II obligate methanotroph, *Methylosinus trichosporium* OB3b, was found to possess PHB up to 30% of total dry weight under various conditions.[42,63,116] This strain possessed both acetoacetate decarboxylase and 3-hydroxybutylase dehydrogenase activities. Acetone accumulation was observed under the same conditions as the decline in PHB. A pathway of acetone production from poly-β-hydroxybutyric acid via 3-hydroxybutyrate and acetoacetate was suggested.[116] The production of PHB was also found in the methanol-grown facultative methylotroph, *Methylobacterium organophilum*.[117] PHB has been considered as a storage polymer for methylotrophic microorganisms. The precise role of the conversion of this storage polymer into acetone as well as of the regulation of PHB synthesis and its subsequent motilization both require further clarification.

B. Energy Generation

The oxidation of methane to carbon dioxide in methanotrophic bacteria appears to occur via a series of two-electron oxidation steps involving the enzymes methane monooxygenase, methanol dehydrogenase, formaldehyde dehydrogenase, and formate dehydrogenase (Figure 7). Formaldehyde occupies a key position in the oxidative pathway as it is either assimilated into cell mass or dissimilated to carbon dioxide to generate energy for growth. Enzymes catalyzing each step have been detected in the cell-free system and have been purified. Reaction mechanisms of these steps have been elucidated at the molecular level.

1. Oxidation of Methane to Methanol

As early as 1959, Leadbetter and Foster[118] presented indirect evidence for the involvement of an oxygenase reaction in methane oxidation and proposed a monooxygenase mechanism:

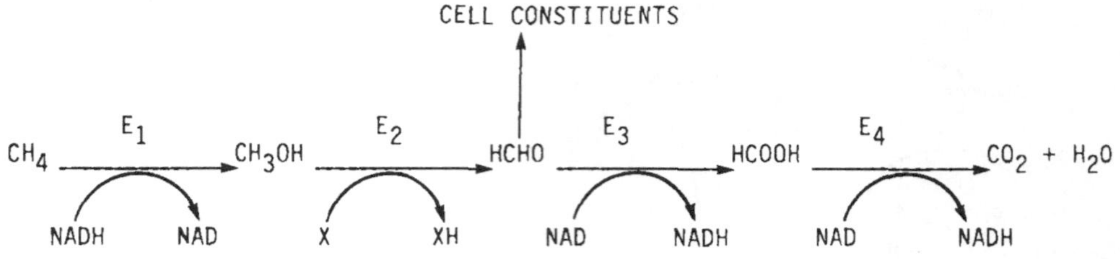

E_1 = METHANE MONOOXYGENASE
E_2 = METHANOL DEHYDROGENASE OR ALCOHOL OXIDASE
E_3 = FORMALDEHYDE DEHYDROGENASE
E_4 = FORMATE DEHYDROGENASE

FIGURE 7. Methane dissimilation pathway.

$$CH_4 + O_2 + XH_2 \rightarrow CH_3OH + H_2O + X$$

They demonstrated that *P. methanica* grown in the presence of ^{18}O-enriched atmosphere incorporated 16 times higher ^{18}O into cell material when methane was carbon source rather than methanol. Direct evidence for oxygenase involvement in the oxidation of methane came, not until 1970, from measurements of ^{18}O incorporation by Higgins and Quayle.[119] They isolated $CH_3{}^{18}OH$ as the product of methane oxidation when suspensions of *P. methanica* or *Methanomonas methanoxidans* were allowed to oxidize methane in $^{18}O_2$-enriched atmosphere. These findings, however, did not exclude the possibility of an unusual dioxygenase mechanism:

$$2CH_4 + O_2 \rightarrow 2\ CH_3OH$$

The subsequent observation of methane-stimulated reduced nicotinamide adenine dinucleotide oxidation catalyzed by extracts of *Methylococcus capsulatus*[52] suggested that the enzyme responsible for this oxygenation is a monooxygenase. Later, in more detailed studies, Ribbons[51] reported methane-stimulated oxygen consumption and NADH oxidation by particulate fractions derived from *M. capsulatus* (Texas strain). The activity resided mainly in large particles which contained intracytoplasmic membranes that are characteristic of methane-utilizing bacteria. These membranes account for 40 to 60% of the total mass of *M. capsulatus*. The stoichiometry of oxygen consumption and NADH oxidation during methane oxidation was consistent with the involvement of a monooxygenase reaction. Formate was detected as the product of methane oxidation. The particulate fractions also catalyzed the oxidation of methanol and formaldehyde with oxygen as the terminal electron acceptor. He also found considerable variability in the methane monooxygenase activities of individual preparations. Many showed very little activity. Methane-dependent NADH oxidation was inhibited by electron transport inhibitors such as azide, cyanide, British anti-Lewisite, antimycin A, and O-phenanthroline. Ferenci[49] and Ferenci et al.[50] reported the isolation of cell-free particulate fractions from *P. methanica* capable of catalyzing NADH-dependent oxygen consumption using methane and carbon monoxide as a substrate. Stoichiometric conversion of carbon monoxide to carbon dioxide involving a monooxygenase reaction was demonstrated and evidence was presented indicating that methane and carbon monoxide oxidations are catalyzed by a single enzyme system. These particulate fractions also catalyzed the oxidation of ethane. Oxidation of methane and carbon monoxide was competitively inhibited by ethane and ammonium chloride. The activity of this system was very unstable; it lost its activity on storage at 0°C for 24 hr. The activity was inhibited by dithiothreitol, reduced glutathione, and cyanide.

Hubley et al.[120] demonstrated that the oxidation of methane by cell suspensions of *Methylosinus trichosporium* was inhibited by metal-binding compounds. They suggested a role for copper in the methane oxygenase on the basis of the pattern of inhibition and the relief of inhibition by added metal ions. The inhibition of metal-chelating or -binding agents on the oxidation of dimethyl ether and methane, but not methanol, by cell suspensions of *Methylococcus capsulatus* and *Methylosinus trichosporium* was also reported.[121] Evidence suggests the involvement of metal-containing enzymes(s) in the initial step of the oxidation of dimethylether and methane.

The methane monooxygenase assay methods used by earlier investigators have relied on indirect enzyme assays to measure the methane-stimulated NADH disappearance spectrophotometrically or the methane-stimulated O_2 disappearance polarographically. There was nodirect assay method used for measuring the disappearance of methane or the accumulation of methanol. The first direct measurement of methane monooxygenase activity was demonstrated by Colby et al.[122] using a soluble derivative of methane, bromomethane. Higgins and Quayle[119] found that a high phosphate concentration (150 m*M*) would inhibit the further oxidation of methanol; the methane monooxygenase activity could then be measured by methanol formation instead of oxygen or NADH consumption.

The first report on partial purification of methane monooxygenase came from Tonge et al. who detected methanol accumulation in the presence of 150 m*M* phosphate in cell-free extracts of *M. trichosporium*.[123,124] All methane-oxidizing activity was located in particulate fractions that were sedimented at 150,000 × g for 90 min, and both NADH and ascorbate were equally effective electron donors. Fractionation of cell-free extracts revealed that methane monooxygenase consists of three components. One is in soluble form and the other two are particulate-bound which can be solubilized by using phospholipase D treatment. The soluble one is a CO-binding cytochrome c with a redox potential of +310 mV, remarkably high for a C-type cytochrome. All three components have been obtained in pure form.[125] Some properties of the enzyme system are summarized in Table 6. The pure methane monooxygenase has a high specific activity and is fairly stable in storage at 0 to 40°C, but is inactivated by freezing. The system is absolutely dependent on cytochrome c_{co} from *M. trichosporium*. Because the monooxygenase is a copper-containing enzyme, inhibition by metal chelating agents is not surprising. The system, however, is also inhibited by 2-mercaptoethanol and dithiothreitol. The purified enzyme system utilized ascorbate or methanol (in the presence of methanol dehydrogenase) as electron donors. The enzyme cannot be linked directly to NADH, however. Tonge et al.[125] suggested the possibility that electrons were derived from NADH via an electron transport chain to the physiological donor. They proposed that in *M. trichosporium* OB3b, ascorbate was probably directly reducing a carbon monoxide-binding cytochrome c which was essential for monooxygenase activity. In addition to methane, the purified monooxygenase system from *M. trichosporium* OB catalyzed the oxidation of ethane, propane, butane, and carbon monoxide.

We demonstrated that resting cell suspensions of methane-grown methanotrophs epoxidized alkenes (C_2 to C_4 tested) to the corresponding 1,2-epoxides[126] and hydroxylated alkanes to their corresponding secondary alcohols and methyl ketones.[29,127,128] Activity was measured by estimating the products of the reactions by gas chromatography. Both expoxidation of alkenes and the hydroxylation of alkanes were inhibited to the same extent by various metal-chelating compounds such as α,α-bipyridyl, 1,10-phenanthroline, potassium cyanide, im-

Table 6
PROPERTIES OF THE METHANE MONOOXYGENASE SYSTEM FROM *METHYLOSINUS TRICHOSPORIUM* OB3b

	Cytochrome c_{co}	Protein 1	Protein 2
Molecular weight	13,000	47,000	9400
Metal ion content	1 atom iron per molecule; variable amount of copper (0.3—0.8 atoms per molecule)	1 atom copper per molecule	none

Molar ratio of components for optimum activity: 1:1:1
Specific activity: 6 μmol/min/mg of protein
pH optimum: 6.9—7.0
Stoichiometry: 5:4:7 (methane utilization to methanol formation to oxygen consumption)
Substrate specificity: active with CH_4, CO, ethane, *n*-propane, *n*-butane
K_m for methane: 66 μM
Electron donor specificity: ascorbate only for pure system
Effect of inhibitors: sensitive to inhibition by CN^-, chelating agents (especially effective copper chelators) SKF 525A, Lilly compounds 18947 and 53325, 2-mercaptoethanol, and dithiothreitol

From Wolfe, R. S. and Higgins, I. J., *Int. Rev. Biochem.*, 21, 267, 1979. With permission.

idazole, and thiourea. Subsequently, both the epoxidation and hydroxylation activities were demonstrated in cell-free systems.[55] Evidence was presented indicating that both reactions are catalyzed by methane monooxygenase activity which resides mainly in the particulate fractions of methane-utilizing bacteria. Details on the epoxidation of alkenes by methanotrophs will be discussed in Chapter 6.

Contrary to these methane monooxygenase activities from particulate fractions, Colby and Dalton reported a soluble methane-oxidizing system from *Methylococcus capsulatus*. (Bath).[129] Soluble methane monooxygenase catalyzed the oxygenation of alkanes, alkenes, ethers, cyclic, alicyclic, and aromatic compounds.[130] They found that only NADH, not ascorbate, would act as an electron donor in this system. Methane monooxygenase activity was not inhibited by most electron transport inhibitors and metal-chelating agents, indicating that an electron transport chain was not involved in the transfer of electrons from NADH to monooxygenase. Acetylene and hydroxyquinoline strongly inhibited the monooxygenase activity. The soluble methane monooxygenase was resolved into three components and purified by ion exchange chromatography, affinity chromatography and molecular size separation.[5,31] All three components were unstable above 0°C. Component A was reasonably stable at 0°C, but components B and C were very unstable unless kept in the presence of phenylmethylsulfonylfluoride and sodium thioglycolate, respectively. However, all components were stable for at least three months if kept below −20°C. Component A has a molecular weight of about 200,000 daltons and subunits of molecular weight 68,000 and 47,000. It contains 2 g-atom of iron and acid-labile sulfide per mole. Upon reduction with sodium dithionite (NADH is not effective), an electron paramagnetic resonance (EPR) signal appears which is enhanced fourfold in the presence of substrate ethylene, suggesting that protein A binds the substrate under reducing conditions. Component C is an iron flavoprotein of molecular weight 44,600 and contains 1 mol of FAD, 2 g-atom of nonheme iron, and 2 mol of acid-labile sulfur per mole of protein.[131] Core extrusion an EPR studies indicated that the component C is of the [2Fe-2S* $(S\text{-}Cys)_4$] class of nonheme iron protein. Component C was directly reduced by NADH under anaerobic conditions, being fully reduced upon addition of 1.2 mol NADH per mole protein, suggesting that component C accepts two electrons per mole of protein. It was reoxidized anaerobically upon the addition of a stoichiometric amount of component A. The other component, component B, is a colorless protein of 15,000 mol wt. At present, its role in the reaction is uncertain. The properties of the component proteins from *M. capsulatus* (Bath) thus appear to be quite different from those reported from *Methylosinus trichosporium* OB3b.[125]

However, Stirling and Dalton[132] reported that *M. trichosporium* OB3b contained soluble methane monooxygenase activity very similar to that from *Methylococcus capsulatus* (Bath). The only effective electron donor for monooxygenase was NAD(P)H and neither ascorbate nor methanol could serve as an electron donor in the presence of soluble carbon monoxide binding cytochrome c or methanol dehydrogenase, respectively. Soluble methane monooxygenase from *Methylosinus trichosporium* OB3b has a broad substrate specificity similar to that obtained from *Methylococcus capsulatus* (Bath). Furthermore, the OB3b methane monooxygenase system is stable at freezing, and insensitive to amytal and cyanide concentration below 1 m*M*. These results are quite different from those reported by Tonge et al.,[124,125] and are in close agreement with those reported by Colby and Dalton[129,131,133] for the *M. capsulatus* (Bath) enzyme. Stirling and Dalton[132] also observed cross-reactivity between components B and C from *Methylococcus capsulatus* (Bath) and component A from *Methylosinus trichosporium* OB3b, suggesting a functional relationship between the two monooxygenases.

More recently, Scott et al.[44] reported that the distribution of methane monooxygenase activities between particulate and soluble fractions of cell-free extracts of *M. trichosporium* OB3b was dependent on growth conditions. Particulate activity was associated with the presence of intracytoplasmic membranes observed only under oxygen-limiting conditions in shake flask cultures. Soluble monooxygenase activity was obtained when the organisms were grown under high oxygen tension in a fermentor. Particulate and soluble activities showed substantially different sensitivities to a range of potential inhibitors. The particulate enzyme was inhibited by metal-chelating agents, thiol reagents, and amytal, whereas the soluble monooxygenase was insensitive to the above inhibitors. In addition, Stanley et al.[287] reported that the intracellular location of methane monooxygenase activity (MMO) in the methanotroph *Methylococcus capsulatus* (Bath) depended primarily on the availability of copper. MMO activity was observed in the particulate fraction of cell extracts under conditions of copper excess but switched to a soluble location in response to copper stress. The two activities could be differentiated by sensitivity to a range of inhibitors and by major changes in the polypeptide banding patterns on denaturing polyacrylamide gels. MMO activity concomitant with the oxidation of ethanol was only observed in cells with particulate MMO activity but could be lost independently in response to copper stress. They also found this copper effect in *Methylosinus trichosporium* OB3b but that some methanotrophs could not overcome copper stress in this way. Both oxygen tension[44,56] and copper effect[287] may account for the previous dispute among researches about the localization of methane monooxygenase.

Previously we reported that *Methylobacterium* sp. CRL-26 grown in shake flasks with methane (methane and air, 1:1 parts by volume) as a carbon source mainly contained particulate methane monooxygenase activity very sensitive to metal-chelating and metal-binding compounds such as 1,10-phenanthroline, 2,2-bipyidyl, thiosemicarbazide, thiourea, and potassium cyanide.[55] Now we have found that *Methylobacterium* sp. CRL-26 obtained by growth in a fermentor with high dissolved oxygen contained mainly soluble methane monooxygenase insensitive to metal-chelating, metal-binding compounds and potassium cyanide. It is possible, therefore, that particulate methane monooxygenase retains a close association with electron transfer protein, and that the inhibition of monooxygenase activity in *Methylobacterium* sp. CRL-26[56] may be due to the inhibition of some electron carrier protein. Soluble methane monooxygenase from *Methylobacterium* sp. CRL-26 catalyzed the NADH- and oxygen-dependent oxygenation of alkanes, alkenes, ethers, cyclic and aromatic compounds. Alkanes (C_1 to C_8 tested) were hydroxylated to the corresponding alcohols.

Table 7
OXIDATION OF ALKENES, CARBON MONOXIDE, ETHERS, CYCLIC AND AROMATIC COMPOUNDS BY SOLUBLE METHANE MONOOXYGENASE FROM *METHYLOBACTERIUM* SP. CRL-26

Substrate	Product	Specific activity (nmol/min/mg of protein)
Ethylene	Ethylene oxide	55
Propylene	Propylene oxide	100
But-1-ene	1,2-Epoxybutane	87
Butadiene	1,2-Epoxybutene	75
Isobutylene	1,2-Epoxyisobutene	95
cis-but-2-ene	*cis*-2,3-Epoxybutane	22
	cis-2-Buten-1-ol	15
trans-but-2-ene	*trans*-2,3-Epoxybutane	25
	trans-2-Buten-1-ol	18
2-Methyl-l-butene	ND	42
2-Methyl-2-butene	ND	16
1-Bromo-1-butene	ND	83
2-Bromo-2-butene	ND	30
Isoprene	1,2-Epoxyisoprene	38
Carbon monoxide	Carbon dioxide	30
Dimethylether	Methanol	25
	Formaldehyde	10
Butylether	ND	34
Cyclohexane	Cyclohexanol	36
Toluene	Benzylalcohol	22
	Cresol	15
Benzene	Phenol	20

Note: ND — Products were not identified.

Both primary and secondary alcohols were produced from the oxidation of C_3 to C_8 alkanes. Among monohalogenated methane derivatives, bromomethane and chloromethane were oxidized more rapidly than fluoromethane. Formaldehyde was detected as the product of chloromethane and bromomethane oxidation. Dichloromethane and trichloromethane were oxidized more slowly than chloromethane. 1-Nitropropane and 1-bromobutane were oxidized more rapidly than were the 2-nitropropane and 2-bromobutane. The branched chain alkane isobutane was oxidized to a mixture of isobutanol and tertiary butanol (Table 7).

The terminal alkenes, ethylene, propylene, 1-butene, and butadiene were oxidized to the corresponding 1,2-epoxides by soluble methane monooxygenase from *Methylobacterium* sp. CRL-26. Internal alkenes such as *trans*-but-2-ene and *cis*-but-2-ene were also oxidized. A mixture of *trans*-2,3-epoxybutane and *trans*-2-but-1-ol was produced from *trans*-but-2-ene indicating the oxidation of the internal double bond and the terminal methyl group. Substituted alkenes such as 2-methyl-1-butene, 2-methyl-2-butene, 1-bromo-1-butene, 2-bromo-2-butene, and branch-chain alkenes isobutylene and isoprene were oxidized (Table 8).

Carbon monoxide, dimethylether, butylether, cyclohexane, toluene, and benzene were oxidized by a soluble methane monooxygenase from *Methylobacterium* sp. CRL-26 (Table 8).

A number of electron donors were unable to replace NAD(P)H as a cofactor for soluble methane monooxygenase from *Methylobacterium* sp. CRL-26. Ascorbate or methanol in the presence of cobinding cytochrome c or methanol dehydrogenase could not replace NADH as an electron donor. Methane monooxygenase from *Methylobacterium* sp. CRL-26 was

Table 8
HYDROXYLATION OF ALKANES BY SOLUBLE METHANE MONOOXYGENASE FROM *METHYLOBACTERIUM* SP. CRL-26

Substrate	Product	Specific activity (nmol/min/mg of protein)
Methane	Methanol	93
Ethane	Ethanol	64
Propane	1-Propanol	15
	2-Propanol	22
Butane	1-Butanol	43
	2-Butanol	25
Pentane	1-Pentanol	21
	2-Pentanol	45
Hexane	1-Hexanol	39
	2-Hexanol	21
Heptane	1-Heptanol	12
	2-Heptanol	50
Octane	1-Octanol	2.5
	2-Octanol	17
Chloromethane	Formaldehyde	44
Dichloromethane	ND	40
Trichloromethane	ND	21
Bromomethane	Formaldehyde	48
Fluoromethane	Formaldehyde	19
Nitromethane	ND	12
Nitroethane	ND	18
1-Nitropropane	ND	50
2-Nitropropane	ND	19
1-Bromobutane	ND	49
2-Bromobutane	ND	10
Isobutane	Isobutanol	34
	tert. butanol	40

resolved into three fractions by ion exchange chromatography. All three fractions were required for maximum activity. Fraction A and B are reddish brown and fraction C is yellow after separation by ion exchange chromatography.

2. *Possible Free Radical Mechanism for Methane Oxidation*

The broad substrate specificity of methane monooxygenase from various organisms, indicates the possible involvement of a free radical mechanism for methane oxidation. Hutchinson et al.[134] proposed a mechanism involving free radicals for methane oxidation by *Methylococcus capsulatus* rather than a monooxygenase system. The process generates a methyl radical from methane by an iron (III) hydroperoxide (Figure 8). The methyl radical can be stabilized by coordination to an iron atom (Figure 9). The iron-methyl reacts with molecular oxygen to produce a methylperoxy iron derivative. The addition of two methyl peroxide radicals with a loss of an oxygen molecule and a further loss or addition of protons generates methanol or formaldehyde. Experimental evidence for radical mechanism is, however, lacking. Nuclear resonance studies of methane monooxygenase in *Methylococcus capsulatus* (Bath) suggested the involvement of an "NIH Shift" mechanism during the hydroxylation of 1-phenylethane by monooxygenase.[135]

3. *Oxidation of Methanol (Methanl Dehydrogenase)*

The NAD(P)$^+$-independent methanol dehydrogenase was first studied in detail in a

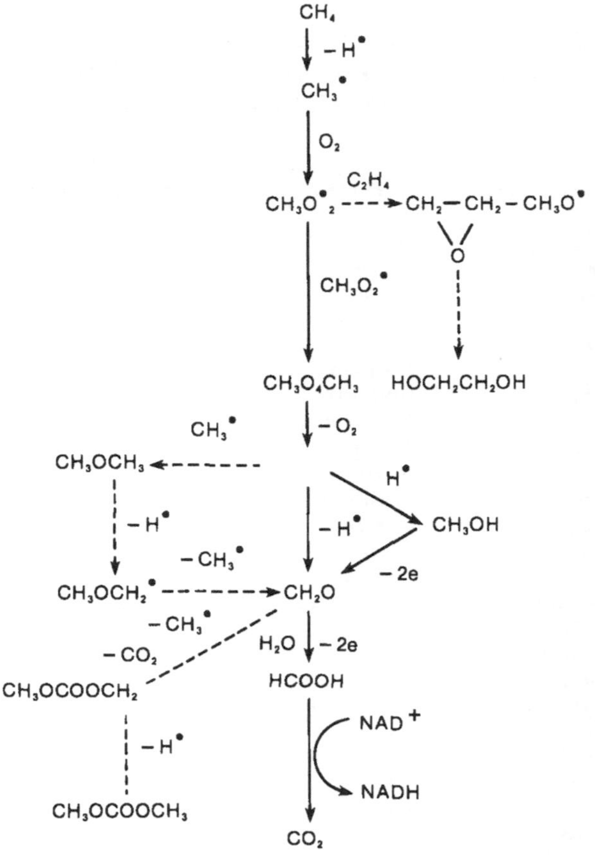

FIGURE 8. Possible pathway for a free radical oxidation of methane. (From Hutchinson, D. W., Whittenbury, R., and Dalton, H., *J. Theor. Biol.*, 58, 325, 1976. With permission.)

facultative methanol-utilizer, *Pseudomonas* M27, by Anthony and Zatman.[136-139] In vitro methanol oxidation can be coupled to phenazine methosulfate indophenol or cytochrome c. Ammonia or methylamine are required as activators. Methanol dehydrogenase from *Methylococcus capsulatus* (Texas) was studied by Patel et al.[140] and was compared with the enzyme from *Pseudomonas* M27. Both enzymes have a wide substrate specificity, oxidizing many primary alcohols, the rate decreasing with increasing chain length. Secondary and tertiary alcohols were not oxidized. The enzyme does not contain any metal ions, has a high pH optimum for activity (pH 9 to 11), and has an absorption peak at 350 nm. Methanol dehydrogenase has been purified from many methanol-utilizing bacteria, and can be roughly divided into four groups according to their properties.[5] Group 1 consists of enzymes that catalyze the oxidation of primary alcohols only, require ammonia or methylamine for activity, and are dimeric proteins with a molecular weight of about 120,000 to 150,000. Enzymes within this group differ in their isoelectric points and amino acid compositions. These enzymes are from *Pseudomonas* M27,[138] *Methylococcus capsulatus* (Texas),[140] *Hyphomicrobium* W C,[141] *Pseudomonas* TP1,[141] *Pseudomonas* W1,[141] *Pseudomonas* 2491,[142] *Hyphomicrobium* X,[143] *Pseudomonas fluorescens* S25 and S50,[144] *Pseudomonas* RJ1,[145] and *Paracoccus denitrificans*.[146] Group 2 are those enzymes from *Pseudomonas* C[147] and *Methylobacterium organophilum*.[148] They resemble those enzyme in group 1 in most respects

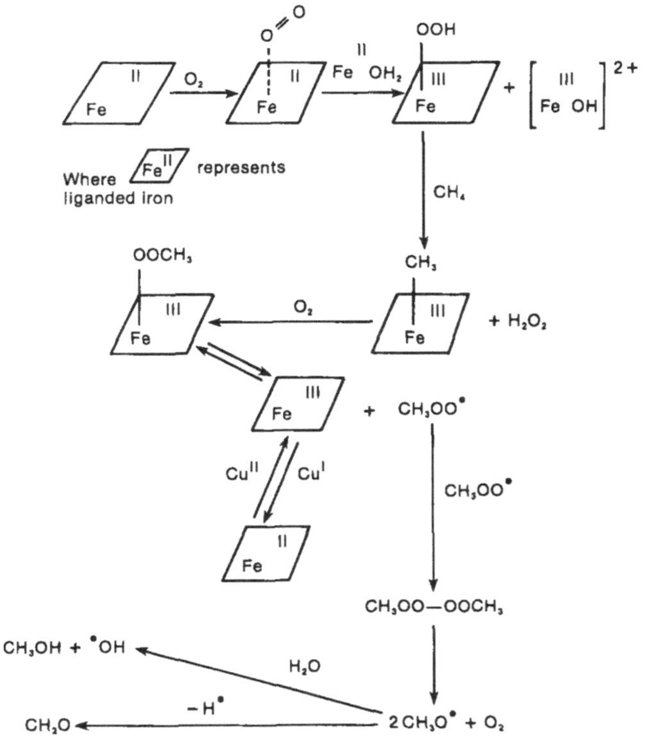

FIGURE 9. Possible role of metal ions in stabilization of free radicals that may involve in methane oxidation. (From Hutchinson, D. W., Whittenbury, R., and Dalton, H., *J. Theor. Biol.*, 58, 325, 1976. With permission.)

except that they also oxidize some secondary alcohols. Group 3 are enzymes from *Methylosinus sporium*[149] and *Methylosinus methanica*.[150] They resemble the group 1 enzymes in many aspects except that they are monomeric with a molecular weight of 60,000. Group 4 consists of enzymes from *Rhodopseudomonas acidophila*.[151] These enzymes differ from group 1 enzymes in their ability to oxidize secondary alcohols, in low activity with methanol, and in their sensitivity to oxygen.

Patel et al.[140] reported the oxidation of formaldehyde by the purified methanol dehydrogenase from *M. capsulatus* (Texas). A dual role for oxidation of methanol and formaldehyde in vivo by this enzyme was proposed. Sperl et al.[141] suggested that the oxidation of formaldehyde by methanol dehydrogenase is due to the hydration of formaldehyde in aqueous solution. Wadzinski and Ribbons[54] purified the methanol dehydrogenase from *M. capsulatus* (Texas) and showed that it is membrane bound in vivo. About 60% of the total methanol-oxidizing activity was found in the membrane fractions. The membrane bound methanol oxidase activity, upon solubilization, was shown to be indistinguishable from soluble methanol dehydrogenase. They propose that the major site for methanol and formaldehyde oxidation lies on the intracytoplasmic membranes where it is coupled to oxygen reduction. Patel et al.[150] compared the immunochemical properties of methanol dehydrogenase from various methane-utilizing bacteria. Immunologically, the methanol dehydrogenase from the type II methane-utilizer is more closely related to those from the facultative methanol-utilizer than from type I methane-utilizing bacteria.

Duine et al.[152] reported the isolation of methanol dehydrogenase from *Hyphomicrobium*

X with a functional coupling to cytochrome c. Oxidized cytochrome c shows a characteristic reduced spectrum on addition of methanol to the complex. Furthermore, this coupling is only functional in anaerobically prepared extracts and is irreversibly destroyed on exposure to oxygen, yielding the conventional, ammonium activated methanol dehydrogenase. There is some evidence that the in vivo electron acceptor for the dehydrogenase is a flavoprotein. Until recently, little was known about the structure of the prosthetic group and the mechanism of action of methanol dehydrogenase. Anthony and Zatman[153] earlier reported that the prosthetic group of methanol dehydrogenase resembled a pteridine derivative in its spectral properties; subsequently, a lumazine derivative was proposed as a prosthetic group. Evidence had accumulated that the coenzyme of methanol dehydrogenase is a novel, nitrogen-containing O-quinone.[152,154,155] Salisbury et al.[156] isolated and crystallized the cofactor of methanol dehydrogenase as fluorescent orange needles and determined its structure by X-ray diffraction analysis. The structure obtained from this analysis was 4,5-dihydro-5-hydroxy-4-oxo-5(2-oxopropyl)-1H-pyrrolo [2,3-f] quinoline-2,7,9-tricarboxylic acid; the trivial name of methoxatin was proposed for the cofactor. Duine et al.[157] using spectral properties and physicochemical technique including NMR spectroscopy, mass spectroscopy, and high pressure liquid chromatography of the isolated cofactor of methanol dehydrogenase reported a similar structure. They designated methanol dehydrogenase as quinoprotein dehydrogenase and proposed the name pyrrolo quinoline quinone (PQQ), since the O-quinone structure is essential for activity. They were able to reconstitute apoenzyme of glucose dehydrogenase, a quinoprotein similar to methanol dehydrogenase, using the cofactor (PQQ) isolated from methanol dehydrogenase.[152] Forrest et al.[158] have proposed a mechanism for the oxidation of methanol involving methoxatin which forms an additional compound with the amino group of a lysine residue (or with ammonia or primary amine activator) at the C_4 position (Figure 10). A 1,4-elimination of water produces the quinone analog in the active system. A 1,4-addition of primary alcohol allows the cyclic rearrangement with the release of the oxidized product. Duine et al.[159] demonstrated that methanol dehydrogenase contains two molecules of prosthetic groups; one is detected as the reduced and the other as the oxidized form. Recently, Mincey et al.[160] reported that approximately 13% of the methanol dehydrogenase cofactor is present in the semiquinone form, and the inactivation of the enzyme leads to concomitant loss of the semiquinone and the enzyme activity. They proposed that upon addition of substrate to the enzyme leads to a two-electron reduction of the semiquinone form of the enzyme. They also suggested the formation of a reduced enzyme and subsequent reoxidation of the reduced enzyme by phenazine methosulfate.

4. *Oxidation of Formaldehyde (Formaldehyde/Aldehyde Dehydrogenase)*

Methylotrophic bacteria contain two different types of formaldehyde (aldehyde) dehydrogenase activity in crude soluble extracts. The first group includes $NAD(P)^+$-linked enzymes that are specific for formaldehyde[112,161-167] as well as nonspecific aldehyde dehydrogenase. Many of the enzymes within this group require glutathione[112,161-163] for their activity. It was suggested that the thiohemiacetal of formaldehyde and glutathione may be the actual substrate for the enzyme.[112] Recently, the $NAD(P)^+$-linked glutathione-independent formaldehyde dehydrogenase has been purified and characterized from a methane-utilizing organism, *Methylococcus capsulatus* (Bath).[167] The enzyme consists of a heat-stable, dialyzable and a heat-sensitive, nondialyzable components. The heat-stable component is trypsin-sensitive and presumed to be a low molecular weight protein or peptide. The heat-sensitive component has a molecular weight of about 115,000 and consists of two equal subunits with a molecular weight of 57,000. The purified formaldehyde dehydrogenase oxidized three other aldehydes with the following Km values: 0.68 m*M* (formaldehyde); 0.075 m*M* (glyoxal); 7.0 m*M* (glycoaldehyde); and 2.0 m*M* (DL-glyceraldehyde). The enzyme has a temperature and pH optima of 45°C and pH 7.2, respectively. Subsequently, Stirling and Dalton[132] reported

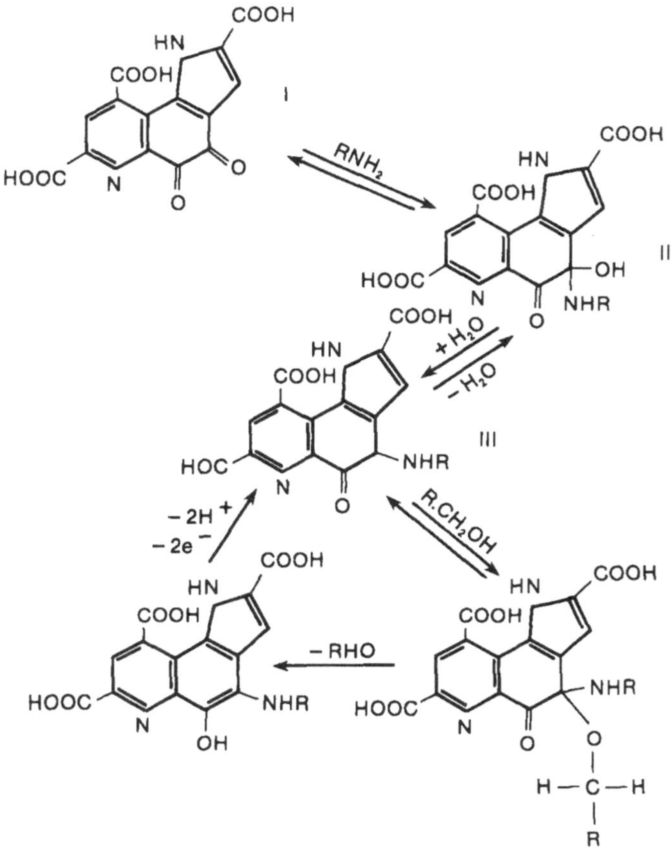

FIGURE 10. Proposed mechanism for the role of methoxatin in methanol oxidation by methanol dehydrogenase. (From Forrest, H. S., Salisbury, S. A., and Kitty, C. G., *Biochem. Biophys. Res. Commun.*, 97, 248, 1980. With permission.)

evidence for a NAD^+-linked enzyme in cell-free extracts of *Methylosinus trichosporium* OB3b.

The second group of formaldehyde dehydrogenase requires an artificial electron acceptor such as phenazine methosulfate (PMS) or 2,6-dichlorophenol indophenol (DCPIP) to assay their activity in vitro, and again, many can utilize various other aldehydes as substrates.[161,163,168-173] The first aldehyde dehydrogenase with broader substrate specificity from the methylotrophic organisms was reported by Johnson and Quayle [161] The partially purified enzyme from the facultative methanol-utilizing organism, *Pseudomonas* AM1, catalyzed the oxidation of straight-chain aldehydes. Aromatic aldehydes were not oxidized. Among various electron acceptors tested, only DCPIP was reduced in the presence of aldehyde. Recently, Patel et al.[171] reported that the purified aldehyde dehydrogenase from the obligate methanol-utilizing organism, *Methylomonas methylovora*, differed from the above enzyme in its requirement for a primary electron acceptor. Among various electron acceptors tested, the enzyme catalyzes the oxidation of aldehydes in the presence of PMS, phenazine ethosulfate, or DCPIP. The purified enzyme also differs from previously described aldehyde dehydrogenases in its ability to catalyze the oxidation of aromatic aldehydes. Evidence shows that this is a heme-containing enzyme.

Another heme-containing PMS-linked aldehyde dehydrogenase was purified from the obligate methane-grown bacterium, *Methylosinus trichosporium* PG.[172] The purified enzyme catalyzes the oxidation of straight-chain aldehydes, aromatic aldehydes, glycoxylate, and glyceraldehyde. Both the purified aldehyde dehydrogenases from an obligate methanol-utilizer, *Methylomonas methylovora*, and from the type II methanotroph; *Methylosinus trichosporium* PG have a molecular weight of 43,000 with two subunits of 22,000 mol wt. Antisera prepared against the purified enzyme from *M. trichosporium* PG are nonspecific; they crossreact with isofunctional enzymes from other methylotrophic bacteria on Ouchterlony double-diffusion plates.

A dye-linked formaldehyde dehydrogenase was also partially purified and characterized from methanol-, methylamine-, or ethanol-grown *Hyphomicrobium* X.[173] The molecular weight of this enzyme is about 83,500. Its optimum pH for the oxidation of formaldehyde was at pH 7.2 using cytochrome c as the electron acceptor and at pH 7.6 when PMS and DCPIP were used. This enzyme catalyzed the oxidation of straight-chain aldehydes (C_1-C_7), glyoxal, glyoxylate, methylglyoxal, glyceraldehyde, and glutaldehyde. Marison and Attwood[173] concluded that this dye-linked enzyme is not induced during growth on C_1 compounds and is a general aldehyde dehydrogenase unlikely to play a major role in the oxidation of formaldehyde.

Johnson and Quayle[161] suggested that formaldehyde is oxidized by a sequence of tetrahydrofolate (THF)-linked enzymes:

$$\text{Formaldehyde} + \text{THF} = N^{5,10}\text{-methylene THF} \quad [1]$$
$$N^{5,10}\text{-Methylene THF} + \text{NADP} = N^{5,10}\text{-methenyl THF} + \text{NADPH}_2 \quad [2]$$
$$N^{5,10}\text{-Methenyl THF} = \text{H}_2\text{O} + N^{10}\text{-formyl THF} \quad [3]$$
$$N^{10}\text{-Formyl THF} + \text{ADP} + \text{Pi} = \text{THF} + \text{formate} + \text{ATP} \quad [4]$$
$$\text{Sum: Formaldehyde} + \text{NADP} + \text{ADP} + \text{Pi} = \text{formate} + \text{NADPH}_2 + \text{ATP} \quad [5]$$

Reaction [1] can occur nonenzymatically. The enzymes catalyzing reactions [2] to [4] have been demonstrated in extracts of methanol-grown *Pseudomonas* AM1.[96]

There are also reports showing that *Pseudomonas* AM1 and *Hyphomicrobium* X can assimilate formate via a sequence of reactions involving tetrahydrofolate derivatives.[91,174] Since the enzymes also catalyze the reversible reaction,[175-177] it seems possible that they could function in the oxidation of formaldehyde to formate. Recently, the activities of enzymes involved in the methylene-tetrahydrofolate pathway were shown to be induced during growth on methanol or methylamine.[178]

Another type of enzyme which catalyzes the oxidation of formaldehyde is the PMS-linked, ammonium ion-dependent methanol dehydrogenase as described previously.[139,140,150] Methanol dehydrogenase contains a quinone derivative as the prosthetic group and does not catalyze the oxidation of G_3 to C_{10} aldehydes.[143,156]

Another route for the oxidation of formaldehyde to CO_2 operating in the RMP pathway bacteria is by the cyclic oxidation shown in Figure 3A (see Section III.A.).

5. *Oxidation of Formate (Formate Dehydrogenase)*

Formate dehydrogenase (EC 1.2.1.2) occurs in animals, higher plants, and microorganisms. Enzyme preparations of high purity have been obtained from anaerobic bacteria such as *Clostridium thermoaceticum*,[179] *C. pasteurianum*[180] and *Escherichia coli*,[181] and from aerobic bacteria such as *P. oxalaticus*.[182] Formate dehydrogenases from yeasts are discussed in Chapter 2. Formate dehydrogenases from bacteria are very labile and are extremely sensitive toward oxygen.[180-182] They have in common a complex chromophore containing metals such as molybdenum, selenium, tungsten, and iron. In methylotrophs, NAD-linked formate dehydrogenase is the last enzyme in the dissimilative pathway of methane or methanol. Formate dehydrogenase from formate-grown *P. oxalaticus*[182] is a complex flavoprotein containing 2 FMN, 18 to 25 nonheme iron atoms and 15 to 20 acid-labile sulfides. The enzyme has a molecular weight of 315,000 consisting of two equal subunits. Recently, an NAD-linked formate dehydrogenase was purified from a methanol-grown methylotroph, *Achromobacter parvulus*.[183] This formate dehydrogenase can be stabilized by -SH compounds, and is insensitive to oxygen. The enzyme retained full activity upon storage at ambient temperature in solution for half a year in the presence of 2-mercaptoethanol or EDTA. Its molecular weight is relatively small, about 80,000. It is composed of two identical or very similar subunits. This enzyme is specific for formate and NAD as electron acceptor. The Michaelis constant was 0.11 m*M* for NAD and 15 m*M* for formate at pH 7.0 and 37°C.

There has been controversy over the nature of the true substrate of formate dehydrogenase. The affinity of formate dehydrogenases for formate is surprisingly low. Km values for formate of a number of formate dehydrogenases of methylotrophs have been reported to vary from 13 to 55 m*M*.[182-184] The formation of S-formylglutathione as the product of NAD-dependent formaldehyde oxidation in *Hansenula polymorpha*[185] implies that during growth of the yeast on methanol, an enzyme should be present which catalyzes the hydrolysis of S-formylglutathione to glutathione and formate in order to supply the formate dehydrogenase with its substrate. Van Dijken et al.[184] found that the formaldehyde-dependent increase in the extinction at 240 nm, resulting from the formation of S-formylglutathione by formaldehyde dehydrogenase, rapidly decreased upon the addition of partially purified formate dehydrogenase, indicating that formate dehydrogenase itself might have a hydrolase activity. However, it seems unlikely that this hydrolase activity of formate dehydrogenase results in the formation of free formate; first, because the hydrolysis of S-formylglutathione by the enzyme was not observed in the absence of NAD and second, because of the finding that the partially purified enzyme has a Km for S-formylglutathione which is approximately 40 times lower than for formate. They suggested that the hydrolysis of S-formylglutathione by the formate dehydrogenase of *H. polymorpha* may lead to the formation of enzyme-bound formate.

Recently, NAD-linked formate dehydrogenase from the methylotrophic bacterium, *Achromobacter parvulus*, was also found to be active with S-formylglutathione as a substrate.[186] The purified enzyme requires the presence of the cofactor NAD for its hydrolytic activity. Evidence obtained led Egorov et al.[186] to conclude that formate dehydrogenase catalyzes the reaction with S-formyl glutathione and NAD.

Formate dehydrogenase from methane-utilizing bacteria has not been purified due to its extreme instability.

6. *Secondary Alcohol Dehydrogenase*

Phenazine methosulfate (PMS)-dependent methanol dehydrogenase has been reported from many methylotrophic bacteria (see above). This enzyme oxidizes primary alcohols from C_1 to C_{10}, but shows either little or no activity toward secondary alcohols (see above). Recently, Hou et al.[29] observed a secondary alcohol-specific dehydrogenase activity in whole cell suspensions and in cell-free extracts of many obligate and facultative methane- and methanol-grown microbes. In cell-free extracts, the dehydrogenase requires a cofactor NAD^+ for its activity. Methyl ketones produced from secondary alcohols are accumulated without further degradation. Although Leadbetter and Foster[187] reported that *n*-alkanes are cooxidized to methyl ketones by *P. methanica*, Hou et al.[29] were the first to show that this is due to the presence of NAD^+-linked secondary alcohol dehydrogenase in methylotrophs. Both primary and secondary alcohols are the products of short-chain *n*-alkane oxidation by methane monooxygenase.[128,130] The production of methyl ketones from *n*-alkanes by cell suspensions of methylotrophs represents two steps. The first step reflects methane monooxygenase activity

Table 9
SUBSTRATES AND STEREOSPECIFICITY OF SECONDARY-ALCOHOL DEHYDROGENASES

Substrates	From *Pseudomonas* sp. ATCC 21439 (%)	From *Pichia* sp. (%)
2-Propanol	85	35
2-Butanol	100	100
2-Pentanol	5	65
2-Hexanol	2	65
3-Pentanol	0	50
3-Hexanol	0	39
3-Octanol	0	27
Methanol to *n*-decanol	0	0
(−)-2-Butanol	140	150
(+)-2-Butanol	40	30
Butane 1,2-diol	0	0
Butane 1,3-diol	2	9
Butane 2,3-diol	2	9
Isobutanol	0	0

and the second reflects secondary alcohol dehydrogenase activity.[127,188] The rate-limiting factor for the production of methyl ketones from *n*-alkanes was found to be the first step, the methane monooxygenase activity.[188]

Secondary alcohol dehydrogenase has also been found in methylotrophic yeasts.[29,189] Both primary alcohol dehydrogenase and secondary alcohol dehydrogenase coexist in most strains of methanol-grown yeasts.[190]

Subsequently, secondary alcohol dehydrogenase was purified from methanol-grown *Pseudomonas* sp. ATCC 21439[191,192] and from methanol-grown *Pichia* sp.[193,194] The enzyme has a molecular weight of 95,000, and consists of two identical subunits and two atoms of zinc per molecule of enzyme protein. It oxidizes secondary alcohols, notably 2-propanol and 2-butanol. Primary alcohols are not oxidized. Secondary alcohol dehydrogenase also catalyzes the reverse reaction, the reduction of methyl ketones to their corresponding 2-alcohols in the presence of NADH. The km values at 25°C in the forward reaction for 2-butanol, (2R)-(−)-butan-2-ol, and NAD, and in the reverse reaction for 2-butanone and NADH are $2.5 \times 10^{-4} M$, $1.6 \times 10^{-4} M$, $1.1 \times 10^{-5} M$, $1.98 \times 10^{-4} M$, and $2.1 \times 10^{-6} M$, respectively. The secondary alcohol dehydrogenase activity was inhibited by metal-chelating agents and by strong thio reagents such as *p*-hydroxymercuribenzoate and 5,5′-dithiobis (2-nitrobenzoic acid). Substrate specificities of secondary alcohol dehydrogenases from both bacteria and yeast are summarized in Table 9. Secondary alcohol dehydrogenase oxidizes preferentially the (−)-2-butanol. This is different from primary alcohol dehydrogenase from bakers' yeast which oxidizes only the (+)-2-butanol. The differences in stereospecificity between primary alcohol dehydrogenase and secondary alcohol dehydrogenase is quite interesting. The stereospecificity of bacterial secondary alcohol dehydrogenase for (−)-2-butanol is perhaps not a surprise when it is considered that two of the groups attached to the asymmetric carbon atoms are methyl and ethyl groups, and that the secondary alcohol dehydrogenase exhibits strict substrate specificity for secondary alcohols. The hydrophobic sites on the secondary alcohol dehydrogenase (or secondary alcohol dehydrogenase-NAD binary complex) can therefore be expected to combine with only a specific alkyl group of 2-butanol. Whether or not the hydroxy group of the substrate reaches to bind with another binding site specific for the hydroxy group depends on the binding between the alkyl groups and the hydrophobic sites. Thus only one enantiomer can be bound with perfect orientation to the active config-

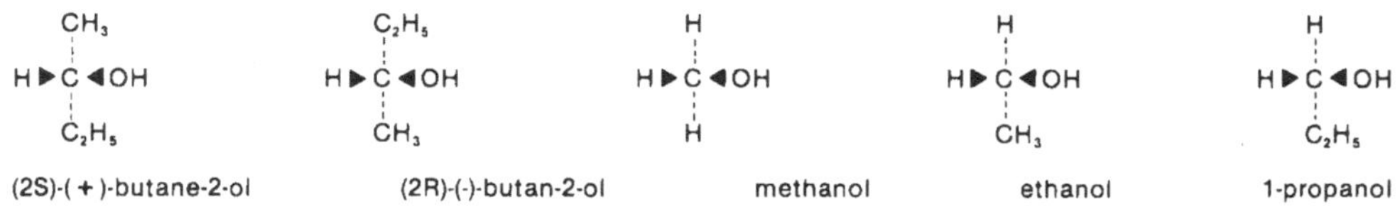

FIGURE 11. Configuration of the enantiomers of 2-butanol, and of methanol, ethanol, and 1-propanol. (From Hou, C. T., Patel, R. N., Barnabe, N., and Marczak, I., *Eur. J. Biochem.*, 119, 359, 1981. With permission.)

uration of secondary alcohol dehydrogenase (or secondary alcohol dehydrogenase-NAD binary complex).

The configurations of enantiomers of 2-butanol, methanol, ethanol, and *n*-propanol are shown in Figure 11. It has been proposed that there is only one alkyl-binding site on the primary alcohol dehydrogenase-NAD binary complex, and that this is essential for the binding of the substrate.[195] In addition, this alkyl group binding site is nonspecific; it binds groups from methyl to decyl. It is clear that secondary-alcohol dehydrogenase requires two binding sites for the two alkyl groups of the substrate.

The physiological significance of this secondary alcohol dehydrogenase in methylotrophs as well as in other gaseous hydrocarbon utilizers is not known. However, possessing this enzyme is of great advantage to the organism since its growth yield, when growing on gaseous alkanes as the sole source of carbon and energy, could be exclusively NAD(P)H-dependent.[196,197] Secondary alcohols are intermediates in the oxidation of *n*-alkanes by either *Methylomonas* or *Mycobacterium*.[196] The methane monooxygenase from *Methylococcus capsulatus* (Bath) and others also oxidizes *n*-alkanes both to primary and secondary alcohols.[55,130] Therefore, the presence of a secondary alcohol dehydrogenase in these gaseous hydrocarbon utilizers seems not a complete surprise. The fact that secondary alcohol dehydrogenase is also present in the methanol-grown cells indicates that the enzyme is not induced by *n*-alkanes.

7. Oxidation of Compounds with Carbon-Carbon Bonds

In spite of their inability to grow on heterotrophic substrates, obligate methanotrophs are capable of oxidizing and incorporating some multicarbon compounds into cell material during growth on C_1 substrates. Wadzinski and Ribbons[198] reported that when *Methanomonas methanooxidans* is supplied with acetate during growth on methane, there is a significant increase in cell yield. Acetate contributes up to 15% of the total cell carbon, indicating considerable assimilation of acetate. The radioactivity from ^{14}C-acetate was incorporated not only into the glutamate and aspartate families of amino acids, but it also appears in those derived from pyruvate, reflecting a functional tricarboxylic acid cycle. On the contrary, *Methylococcus capsulatus* shows restricted incorporation of ^{14}C-acetate into lipid and only four amino acids (glutamate, proline, arginine, and leucine), due to the incomplete tricarboxylic acid cycle in this species.[67,199] Similar results were obtained with ^{14}C-ethanol, suggesting initial oxidation of ethanol to acetate.

Foster[200] discovered that *Pseudomonas methanica* oxidized ethane, propane, and butane while the organism utilized its growth substrate, methane, and termed it as cooxidation. Over the years, the concept of cooxidation has become somewhat confusing. The confusion is due to the extension by more than one author of the original definition of cooxidation to cover the metabolism of nongrowth substrates in the absence of a true growth substrate. Hydrocarbon cooxidation has been thoroughly reviewed by Perry.[201] *Mycobacterium vaccae* was reported to cooxidize many cycloparaffins to the homologous cycloalkanones during propane utilization. Quayle[202] argued for a return to the original terminology and adoption of the term, stationary metabolism, to describe metabolism of nongrowth substrates in the absence of growth substrates.

The discovery of broad substrate specificity of methane monooxygenase[29,126-128,130,188,203] has prompted considerable research effort and much speculation over the metabolic significance of these reactions to methylotrophs. Resting-cell suspensions of methane-grown cultures oxidize gaseous alkenes to their corresponding 1,2-epoxides, and alkanes to their homologous primary and secondary alcohols. They also oxidize a wide variety of substrates including alkanes, alkenes, ethers, and substituted methane derivatives in the absence of methane as a growth substrate. Since there is no cosubstrate involved in these reactions, Hou[204] proposed to use a term, stationary transformations, in order to distinguish them from the widely abused cooxidation. Stirling and Dalton[205,206] defined the oxidation of nongrowth substrates catalyzed by methane monooxygenase and subsequent enzymes as fortuitous, suggesting that these activities are a consequence of the mechanistic requirement for C_1 oxidation. In contrast, Higgins et al.[207,208] argued that because carbon (acetate) and energy (NADH and ATP) can be derived from the oxidation of some environmentally important hydrocarbons this metabolism should be termed supplementary. However, there is no firm evidence to support either the fortuitous or supplementary functions.

8. Electron Transfer and Energy Transducing Systems

Higgins[209] and Wolfe and Higgins[9] gave thorough reviews on electron transfer and energy transducing systems in methanotrophs, although current knowledge in these areas is limited. There are several reports on electron transport components in methanotrophs. Ribbons et al.[1] reported the presence of a- and c-type cytochromes in *Methylococcus capsulatus* but could not detect any b-type cytochromes. They suggested that the obligate dependence of some microorganisms on methane or methanol may be due to their having electron transport and energy transducing systems different from common heterotrophs. Similar results were reported by Davey and Milton[210] who examined whole organisms of *Methylomonas albus* and *Methylosinus trichosporium* and concluded that these bacteria contained cytochrome o in addition to the a- and c-type cytochromes. Weaver and Dugan[211] confirmed the presence of cytochromes a and c in *M. trichosporium* and showed that these cytochromes are located in the membrane fractions together with ATPase activity. Monosov and Netrusov[212] obtained evidence for the presence of a-, b-, and c-type cytochromes in both *M. trichosporium* and *Methylomonas agile*.

Tonge, Higgins and co-workers[213,214] carried out a comparative study of the cytochrome systems of five facultative methylotrophs and methanotrophs, including *Methylosinus trichosporium* and *P. methanica*. They found that the cytochrome components in all five cases were closely similar. The concentrations of particle-bound cytochromes a, b, and c were low, and the concentration of soluble carbon monoxide-binding cytochrome c was high in each case. They could not detect the presence of cytochrome o and suggested that the cytochrome o reported previously[210] was due to this soluble cytochrome c. They indicated that carbon monoxide binding is not usually associated with c-type cytochrome and that it may have an oxidase or oxygenase function for this cytochrome in methylotrophs. They also found different cytochrome complements in a facultative methylotroph when grown on heterotrophic substrates and on methanol. Growth on heterotrophic substrates gave massive increases in the amounts of membrane-bound cytochrome aa_3, b, and c, and the synthesis of some cytochrome a_2 which is not present in methanol-grown cells. This was accomplished by a decrease in the concentration of soluble cytochrome c.

The presence of cytochrome aa_3 and the carbon monoxide-binding cytochrome c in *M. trichosporium* and *P. methanica* was reported using inhibitor studies.[124,125,215,216] Widdowson and Anthony[217] showed the presence of a cytochrome aa_3, a b-type cytochrome, and a co-binding cytochrome c in *P.* AM1. The purified cytochrome c (redox potential 0.26V) of *P.* AM1 was not susceptible to oxidation by molecular oxygen. Studies with whole cells and a cytochrome c-deficient mutant have demonstrated that cytochrome b (redox potential

0.009V) is the first cytochrome in the electron transport chain for the oxidation of all substrates except methanol whose oxidation does not involve this cytochrome.

Drabikowska[218] reported the presence of ubiquinone-8, a cytochrome b and two cytochrome c (c-552.5 and c-549) in the particulate fraction of *Methylomonas* Pl1. The merging point for electron transport from NADH dehydrogenase and formaldehyde dehydrogenase is suggested to be at the level of ubiquinone. A soluble cytochrome c (18,000 mol wt) was purified from *M. methanica.*[219] Immunological properties of cytochrome c from *M. methanica* are similar to those of cytochrome c from various obligate methane-utilizing bacteria.

Two types of soluble cytochrome c have been extensively studied in the obligate methylotroph *Methylophilus methylotrophus*[220] and the facultative methylotroph *P.* AM1.[221] At least one is involved in electron transport from the quinoprotein methanol dehydrogenase.[222,223] These two types of cytochrome c have markedly different isoelectric points, and these have been used as the basis of a provisional terminology: cytochrome c_L has a low isoelectric point and a high molecular weight (about 20,000), whereas cytochrome c_H has a high isoelectric point and a low molecular weight (8500 to 11,000). Free thiol groups were found probably not to be involved in the unusually rapid autoreduction of these cytochromes c at high pH.[224] The axial ligands to the heme iron, histidine and methionine, are the same as in other low-spin cytochrome c. The methionine ligand is displaced at high pH by an alternative strong-field ligand. This displacement does not occur on the reduction of cytochrome c_L by methanol dehydrogenase, but this does not rule out the possibility that the auto reduction mechanism is involved in the interaction of the dehydrogenase and cytochrome c.

In the chemiosmotic theory of oxidative phosphorylation, Mitchell[225] postulates that a transmembrane proton gradient is generated as a result of electron flow down the respiratory chain and that this gradient can drive ATP synthesis such that 1 mol of ATP is synthesized per 2-g-equiv. H^+ translocated. Therefore, the P:O ratio can be calculated from measurements of the $^4H^+$:O ratio (g-equiv. H^+ translocated per g-atom O consumed). Tonge et al.[226] reported a P:O ratio of one for all substrates tested (methane, methanol, formaldehyde, formate, and ascorbate) for *Methylosinus trichosporium*. Netrusov et al.[227] showed that the membrane fraction of *P.* sp 2 cells grown of methanol has the capacity for respiration and ATP synthesis in the presence of formate, methanol, succinate, ascorbate, NADH, or reduced cytochrome c. This synthesis is coupled with the function of the electron transport chain. The oxidation of one molecule of NADH, formate or succinate (P to O ratio = 1.54:2.0) generates more energy (ATP) than the oxidation of one molecule of methanol, reduced cytochrome c or ascorbate (P to O ratio = 0.34:0.54).

O'Keefe and Anthony[228] studied the role of cytochrome c in *P.* AM1 by measuring the stoichiometry of proton translocation driven by respiration of endogenous or added substrates. The maximum $\rightarrow H^+$ to O ratio was about four. They tentatively proposed that the electron transport chain for NADH oxidation in *P.* AM1 is arranged such that the NADH-ubiquinone oxidoreductase forms one proton-translocating segment and that the second segment consists of ubiquinone and cytochromes b and aa_3. The cytochrome c appears to be essential only for respiration and proton translocation from methanol. Dawson and Jones[229] determined the stoichiometry of respiration-linked proton translocation in *Methylophilus methylotrophus*. They showed that $6H^+/O$ are translocated during NADH oxidation and that $2H^+/O$ are translocated during the oxidation of methanol to formaldehyde. It is proposed that the respiratory chain in *M. methylotrophus* is arranged such that there are three sites of energy conservation for NADH oxidation, each translocating $2H^+$ and each linked to the synthesis of one molecule of ATP. Only the third site of energy conservation is involved in methanol oxidation.

9. Oxidation of Methylamine

There are two routes for the oxidation of methylamine in methylotrophic bacteria. The direct route is catalyzed by methylamine dehydrogenase. The second route involves *N*-methylglutamate as an intermediate. The soluble methylamine dehydrogenase oxidizes primary aliphatic amines and diamines to the corresponding aldehydes using phenazine methosulfate as an electron acceptor.

$$CH_3 - NH_3^+ + PMS + H_2O \rightarrow HCHO + NH_4^+ + PMSH_2$$

The natural electron acceptor may be a cytochrome c.[230] Metha[231] partially purified methylamine dehydrogenase from the obligate methylotroph *Methylomonas methylovora*. The enzyme from *M. methylovora* is somewhat similar in its heat stability, pH optima, and inhibition pattern to the amine dehydrogenase of the facultative methylotroph *Pseudomonas* AM1,[232] but it differs from AM1 enzyme in its substrate specificity and metal ion sensitivity. The enzyme from *M. methylovora* was more specific for methylamine (100%) than for ethylamine (16%) and *n*-propylamine (2%). The amine dehydrogenase from strain AM1, however, showed broader substrate specificity with methylamine (100%), ethylamine (98.5%), *n*-propylamine (127%), and *n*-butylamine (100%). The enzyme purified from *P*. sp. J[230] is essentially similar in properties to the enzyme obtained from *P*. AM1. They have a molecular weight of 105,000 with the two kinds of subunits of 40,000 and 13,000 mol wt. The enzymes contain no phosphorous, iron or copper. They have absorption maxima at 278, 330, 430, and 460 nm (shoulder). Only primary amines, not secondary or tertiary amines, are oxidized. Phenazine methosulfate is the most effective electron acceptor; oxygen is ineffective. Matsumoto and co-workers[230,233-235] have made considerable progress on the quaternary structure of methylamine dehydrogenase from *P*. sp. J. It contains two kinds of subunits, light (13,000) and heavy (40,000) in a $\alpha_2\beta_2$ configuration. Coexistence of both of the subunits is essential for the restoration of the enzyme activity. The prosthetic group, assumed to be a derivative of pyridoxal, is bound to the light subunit. The $\alpha_2\beta_2$ arrangement of the subunits was confirmed by using a cross-linking agent, dimethylsuberimidate. The enzyme is possibly assembled around a core consisting of two heavy subunits.

The second route for methylamine oxidation involves two enzymes, *N*-methylamine synthase (EC 2.1.1.21) and *N*-methylglutamate dehydrogenase (EC 1.5.99.5), which catalyze the following reactions:

$$\text{glutamate} + \text{methylamine} \rightarrow N\text{-methylglutamate} + NH_3$$

$$N\text{-methylglutamate} + X + H_2O \rightarrow XH_2 + HCHO + \text{glutamate}$$

Pollock and Hersh[236] purified *N*-methylglutamate synthase from *P*. MA. Hersh et al.[237] subsequently identified *N*-methylglutamate dehydrogenase activity in the particulate fraction of the same microorganism. *N*-Methylglutamate dehydrogenase has been partially purified from *P* MA[238] and purified from *P. aminovorans*.[239,240] Both enzymes use 2,6-dichlorophenol-indophol as in vitro electron acceptor. The enzyme from *P. aminovorans* has a molecular weight of about 550,000 and is possibly a tetramer.[240] NAD-linked *N*-methylglutamate dehydrogenase has been found in *P. methylica*[241] and *Hyphomicrobium vulgare* ZN.[242]

There is evidence for two other mechanisms involving methylamine metabolism. The first one produces *N*-methylamine by *N*-methylalanine dehydrogenase.[243] The second one produces γ-glutamylmethylamide by γ-glutamylmethylamide synthetase (EC 6.3.4.12). The enzyme has been found in *P. oleovorans*,[244] *P*. MA,[245] *P*. MS,[246] *H*. ZV,[247] *H*. X,[248] and *P. aminovorans*.[249]

10. Oxidation of Dimethylamine and Trimethylamine

The initial catabolism of trimethylamine in methylotrophs may proceed by one of two

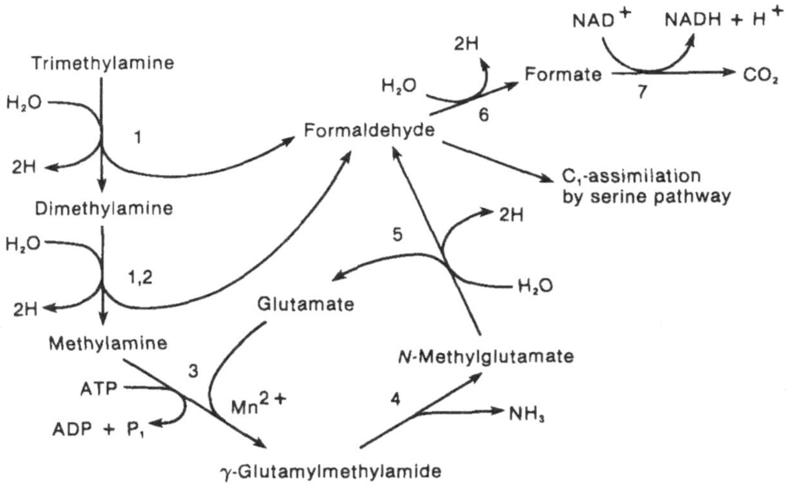

FIGURE 12. Proposed scheme for trimethylamine, dimethylamine, and methylamine metabolism in *Hyphomicrobium* X. The enzymes of the pathway are (1) trimethylamine dehydrogenase (EC 1.5.99.7); (2) dimethylamine dehydrogenase; (3) γ-glutamylmethylamide synthetase (EC 6.3.4.12); (4) enzyme not yet characterized; (5) *N*-methylglutamate dehydrogenase (EC 1.5.99.5); (6) formaldehyde dehydrogenase; (7) formate dehydrogenase (EC 1.2.1.2). (From Meiberg, J. B. M. and Harder, W., *J. Gen. Microbiol.*, 110, 143, 1979. With permission.)

different pathways. The first route involves the direct demethylation to dimethylamine and formaldehyde, catalyzed by trimethylamine dehydrogenase (EC 1.5.99.7).[250]

$$(CH_3)_3NH^+ + \text{flavoprotein} + H_2O \rightarrow (CH_3)_2NH_2^+ + HCHO + \begin{matrix}\text{reduced}\\ \text{flavoprotein}\end{matrix}$$

In the second pathway, the initial attack is an oxygenation of trimethylamine to trimethylamine *N*-oxide mediated by trimethylamine monooxygenase.[251] The *N*-oxide is subsequently demethylated by trimethylamine *N*-oxide aldolase (demethylase) to dimethylamine and formaldehyde.[252,253]

$$(CH_3)_3NH^+ + O_2 + NADPH + H^+ \rightarrow (CH_3)_3NO + H^+ + H_2O + NAD(P)^+$$

$$(CH_3)_3NO + H^+ \rightarrow (CH_3)_2NH_2^+ + HCHO$$

Trimethylamine monooxygenase is very unstable and has been purified about sixfold from extracts of *P. aminovorans*.[236] Trimethylamine *N*-oxide aldolase has been purified from *Bacillus* PM6[252] and *P. aminovorans*.[253] The enzyme has a molecular weight of 45,000 daltons and contains no prosthetic group. Trimethylamine dehydrogenase purified from bacteria 4B6 and W3A1 catalyzes the oxidative *N*-dealkylation of trimethylamine, diethylamine, and a very few other tertiary methyl- and ethyl amines using phenazine methosulfate as in vitro electron acceptor.[254,255] The natural electron acceptor is now known to be a flavoprotein.[256] The molecular weight of the enzyme is 146 to 161 × 10^3 daltons and contains nonidentical subunits of molecular weight 70 to 80 × 10^3 daltons.[250,254,255] Meiberg and Harder[248] investigated the enzymes involved in the metabolism of trimethylamine and dimethylamine by *H.* strain X under both aerobic and anaerobic conditions. On the basis of the results obtained, they propose a pathway for the metabolism of methylated amines in *H.* X (Figure 12). Trimethylamine is oxidized by phenazine methosulfate-linked dehydrogenase which also displays some activity towards dimethylamine.[254,257] The dimethylamine is further oxidized by another dye-linked dehydrogenase which does not show activity towards trimethylamine. Both enzymes are present during growth on dimethylamine and trimethylamine. In *H.* X, methylamine is oxidized via γ-glutamylmethylamide and N-methylglutamate. The formaldehyde generated in the above reactions can either be oxidized to CO_2 via formate or assimilated via the icl$^-$ serine pathway.

Recently, Steenkamp and colleagues[255,256,258-262] have studied trimethylamine dehydrogenase in detail by using a variety of sophisticated physical and chemical techniques, including adsorbance, fluorescence, EPR and NMR spectroscopy, and rapid-freeze EPR. Trimethylamine dehydrogenase from bacterium W_3A was found to contain a novel covalently bound flavin coenzyme, 6-S-cysteinyl-FMN[259,260] and also a tetrameric [4Fe-4S] cluster as redox active groups.[261,262] Kinetic studies by the stopped-flow and freeze-quench techniques indicated that the FMN moiety is rapidly reduced to $FMNH_2$ upon mixing with substrate, whereas the transfer of an electron from $FMNH_2$ to the Fe_4S_4 cluster proceeded much more slowly.[256,261] More recently, Steenkamp and Beinert[263] showed that the trimethylamine dehydrogenase of bacterium W_3A_1 is reduced with the formation of a triplet state in which two electrons, derived from the substrate, are distributed between the Fe_4S_4 cluster and 6-S-cysteinyl-FMN semiquinone. They also showed that the substrate-reduced forms of the trimethylamine and dimethylamine dehydrogenases of *H.* X exist in the triplet state, which arise through interaction of a reduced $[4Fe\text{-}4S]_r$ cluster and flavosemiquinone. The EPR signals differ between the trimethylamine dehydrogenases of bacterium W_3A_1 and *H.* X.[264] The results obtained indicate that the transfer of an electron to the [4Fe-4S] cluster is an obligatory step in the catalysis and suggest that the transfer of electrons from these enzymes to electron acceptors is mediated solely through the [4Fe-4S] cluster.

In aerobic methylotrophic microorganisms, the oxidation of dimethylamine to methylamine and formaldehyde involves a secondary amine monooxygenase. This enzyme has been found in *P. aminovorans*,[265-267] *Bacillus* PM 6,[252] obligate and facultative methylotrophs,[166] and Hyphomicrobia.[242,268] During anaerobic growth of *H.* X on trimethylamine and dimethylamine in the presence of nitrate, however, dimethylamine is oxidized by a dimethylamine dehydrogenase.[248] This is a new enzyme. Previous reports on dimethylamine dehydrogenase activity refers to the ability of purified trimethylamine dehydrogenase to oxidize the substrate at low rates.[248,254] The dimethylamine dehydrogenase from *H.* X grown anaerobically on dimethylamine was purified 15-fold, which was free from trimethylamine dehydrogenase.[269] The molecular weight of the enzyme was 176,000 daltons. It contained two identical subunits of 91,000 daltons. The absorption spectrum showed a maximum at 441 nm. Reduction of the enzyme with dimethylamine produced a new absorption maximum at 356 nm, while the absorption at 441 nm decreased. The pH optimum for the oxidation of dimethylamine was 8.1. The enzyme was absolutely specific towards secondary amines. Phenazine methosulfate served as an artificial electron acceptor. The apparent Km of the enzyme for dimethylamine at pH 7.7 was 15.6 ± 1.6 μM. Steenkamp[270] has identified the prosthetic groups of the dimethylamine dehydrogenase from *H.* X. The enzyme contained a [4Fe-4S] cluster and a covalently bound yellow coenzyme. The coenzyme was identified as 6-S-cysteinyl-FMN, similar to that of trimethylamine dehydrogenase.

C. Anaerobic Methane Oxidation

Evidence has been obtained for the anaerobic oxidation of methane, which couples with reduction of sulfate to hydrogen sulfide.[271] These bacteria were isolated from the sediment surface of Lake Mendota, Wis. They do not incorporate $^{14}CH_4$ into cell material; instead, they incorporate carbon from acetate, lactate, or yeast extract. The biochemistry of this anaerobic methane oxidation is not known.

Zeikus and co-workers in their studies on one-carbon compound metabolism in anaerobic methanogens have demonstrated that these bacteria can grow on methanol, H_2/CO_2, CO, or methylamine as sole sources of carbon and energy.[272,273] They also isolated and characterized a new methylotrophic, acidogenic anaerobe.[274] The name *Butyribacterium methylotrophicum* sp. nov. is proposed for this so-called Marburg strain. The strain is a mesophilic, Gram-positive, nonmotile, pleomorphic rod. Cell morphology varies from single or paired straight rods to rudimentary branched rods, club-shaped cells, or oval refractile cells. Electron micrographs of thin sections revealed a thick monolayered cell wall and an atypical spore structure. The biochemistry of this methylotrophic, acidogenic anaerobe is also not known.

D. Nitrogen Metabolism

The ability of many methanotrophs to fix dinitrogen in addition to using nitrate or ammonia as a nitrogen source is known.[2] Dalton and Whittenbury[30] studied nitrogen metabolism in *Methylococcus capsulatus* (Bath). They detected that glutamate dehydrogenase activity is higher in ammonia-grown organisms. They also detected that glutamine synthetase is more active when the organsm has been grown under nitrogen-fixing conditions. A different strain of *M. capsulatus*, however does not possess glutamate dehydrogenase; instead, it has a glutamine synthetase-glutamine 2-oxoglutarate aminotransferase system.[275,276] Shishkina and Trotsenko[32] studied the pathways of ammonia assimilation in seven obligate methane utilizers including both type I (*Methylomonas methanica, Methylobacter bovis, M. vinelandii*, and *Methylococcus capsulatus*), and type II organisms (*Methylosinus trichosporium, Methylocystis fuscus*, and *M. methanolicus*). They found that the type I strains tested, except *Methylococcus capsulatus* (Texas), had glutamate (or alanine) dehydrogenase and glutamine-synthetase activities. Type II strains tested had the glutamine synthetase-glutamine 2-oxoglutarate aminotransferase system. They also found that enzyme activities varied depending on the nitrogen source used.

Ferenci et al.[50] showed that ammonia is a competitive inhibitor of both methane and carbon monoxide oxidation by *Pseudomonas methanica*, suggesting that ammonia is possibly oxidized by the methane monooxygenase system. O'Neill and Wilkinson[277] have studied ammonia oxidation by methanotrophs, and suggest on the basis of pH effects that NH_3 rather than NH_4^+ interacts with the enzyme. In addition, there is kinetic evidence that ammonia and methane bind to the same site in methane monooxygenase. Hydroxylamine was detected as an oxidation product of ammonia by *Methylococcus capsulatus* (Bath)[278] and by *Methylosinus trichosporium* OB3b.[203] Recently, Sokolov et al.[279] compared the enzyme systems of methanotrophs that oxidize hydroxylamine and methanol. They found a specific hydroxylamine: cytochrome c oxidoreductase, distinct from methanol dehydrogenase, in *Methylococcus thermophilus*.

The first evidence for $^{15}N_2$-fixation by methanotrophs was reported by Coty[280] and later confirmed by deBont and Mulder[31] with a pure culture of *Methylosinus* strain 41. The nitrogen fixation is accompanied by hydrogenase activity which can support acetylene reduction.[281] To detect nitrogenase activity in methane utilizers by measuring ethylene formation from acetylene is generally difficult. deBont and Mulder[31] showed that ethylene was cooxidized in the presence of methane. They suggested that this may be the reason that the product ethylene cannot be measured. Dalton and Whittenbury[30] however, were able to measure acetylene reduction by organisms grown on methanol. It is known that acetylene is a potent inhibitor of methane monooxygenase[30,282-284] and that its presence may therefore preclude methane oxidation from supplying the reducing power required for nitrogenase activity. It is now believed that this inhibitory effect is the cause of the failure in detecting acetylene reduction. Nitrogenase activity in methanotrophs can be measured using the acetylene reduction assay under low oxygen tension and in the presence of a source of reducing power for the enzyme, such as methanol, ethanol, hydrogen, formaldehyde, or formate, but not methane.[30,281,284] It is also possible to measure nitrogenase activity in the presence of methane by using the reduction of nitrous oxide to dinitrogen as the assay method.[30,284] The nitrogenase system of methanotrophs is also oxygen sensitive.

E. The Nature of Obligate Methanotrophy

Obligate methanotrophs are unable to grow on compounds having a carbon-carbon bond. A variety of reasons for this obligateness has been proposed. One of them is the lack of a complete TCA cycle and absence of 2-oxoglutarate dehydrogenase in type I organisms which use ribulose monophosphate cycle for formaldehyde fixation. However, this is unlikely to be the sole basis of obligateness in all methanotrophs because obligate type II methanotrophs using the serine pathway for formaldehyde assimilation do possess all the enzymes of the TCA cycle. Type II obligate methanotrophs lack hexokinase and 6-phosphogluconate dehydrogenase. The lack of these two enzymes impairs only the growth of these organisms on sugars. Therefore, the cause of the obligateness in type II methanotrophs is also not clear.

In *Pseudomonas* AM1, the lack of 2-oxoglutarate dehydrogenase alone has been shown to be the only reason for C_1 obligateness.[285] A mutant of *P.* AM1 lacking this enzyme could only grow on C_1 compounds. The mutant had all the other TCA cycle enzymes and the cytochrome levels were unchanged in the mutant. Two revertants, both of which regained 2-oxoglutarate dehydrogenase activity, also regained the ability to oxidize and grow on the same substrates as wild-type bacteria.

Other possible reasons for the nature of obligate methanotrophy are proposed[9] which include the lack of suitable energy-transducing systems for heterotrophic substrates (see Section III.B.8), the accumulation of toxic metabolites from heterotrophic substrates, and the lack of transcriptional control. The presence of NADH oxidase in all obligate methanotrophs is known. Therefore, it is likely that the last two possibilities are responsible in some cases for the failure of obligate methylotrophs to grow on substrates having a carbon-carbon bond. Attempts have been made, although not successful, to grow *Methylomonas methanooxidans* on glucose in a dialysis culture.[9] Higgins[9] concluded that the lack of 2-oxoglutarate dehydrogenase in type I methanotrophs may be adequate to account for their obligate nature. The reason for this obligateness in type II organisms is less clear.

Recently, Shishkina and Trotsenko[286] found multiple enzymic lesions rather than a single enzymic lesion[275] in obligate methanotrophic bacteria. They studied pathways of pyruvate and acetate metabolism in two type I and two type II obligate methanotrophs. Although both substrates are utilized by these bacteria, the channeling of them into major energetic and biosynthetic pathways seems to be very limited. In *Methylomonas methanica* and *Methylobacter bovis* (type I organisms), pyruvate and acetate can enter both the oxidative branch of the TCA cycle through acetyl-CoA and, to a lesser extent, its reductive branch after carboxylation by malic enzyme. However, due to the absence of α-ketoglutarate dehydrogenase, pyruvate and acetate cannot be completely oxidized to serve as an energy source. In addition, these bacteria do not possess the activities of glyoxylate shunt specific enzymes; isocitrate lyase and malate synthase. They attributed these metabolic defects to the C_1 compound obligateness of these organisms. In *Methylosinus trichosporium* and *Methylocystis minimus* (type II organisms), the metabolism of pyruvate via acetyl-CoA is of minor importance due to the extremely low activity of pyruvate dehydrogenase. Isocitrate lyase and malate synthase are also absent in these organisms. In addition, these organisms lack phosphoenol pyruvate synthase, pyruvate phosphate dikinase, pyruvate carboxylase, and phosphoenol pyruvate-carboxykinase. Thus, in type II obligate methanotrophs, all known gluconeogenic pathways from pyruvate and acetate are blocked. This explains the nature of C_1 obligateness of type II organisms.

Despite the inability to grow on heterotrophic substrates, obligate methanotrophs can utilize and incorporate some of these multiple carbon compounds into cell material during growth on C_1 compounds (see Section III.B.).

REFERENCES

1. **Ribbons, D. W., Harrison, J. E., and Wadzinski, A. M.,** Metabolism of single carbon compounds, *Ann. Rev. Microbiol.*, 24, 135, 1970.
2. **Quayle, J. R.,** The metabolism of one-carbon compounds by microorganisms. *Adv. Microb. Physiol.*, 7, 119, 1972.
3. **Quayle, J. R.** Aspects of the regulation of methylotrophic metabolism, *FEBS Lett.*, 117, K6, 1980.
4. **Anthony, C.,** The biochemistry of methylotrophic bacteria, *Sci. Prog. (London)*, 62, 167, 1975.
5. **Colby, J., Dalton, H., and Whittenbury, R.,** Biological and biochemical aspects of microbial growth on C_1 compounds, *Ann. Rev. Microbiol.*, 33, 481, 1979.
6. **Colby, J. and Zatman, L. J.,** Hexose phosphate synthase and tricarboxylic acid cycle enzymes in bacterium 4B6, an obligate methylotroph, *Biochem. J.*, 128, 1373, 1972.
7. **Hanson, R. S.,** Ecology and diversity of methylotrophic organisms, *Adv. Appl. Microbiol.*, 26, 3, 1980.
8. **Whittenbury, R., Phillips, K. C., and Wilkinson, J. F.,** Enrichment, isolation and some properties of methane-utilizing bacteria, *J. Gen. Microbiol.*, 61, 205, 1970.
9. **Wolfe, R. S. and Higgins, I. J.,** Microbial biochemistry of methane — a study in contrasts, *Int. Rev. Biochem.*, 21, 267, 1979.
10. **Higgins, I. J., Best, D. J., Hammond, R. C., and Scott, D.,** Methane-oxidizing microorganisms, *Microbiol. Rev.*, 45, 556, 1981.
11. **Sohngen, N. L.,** Uber Bakterien, welche methan als kohlenstoffnahrung und Energiequelle Gebrauchen, *Centr. Bakt. Parasitenk.*, 15, 513, 1906.
12. **Orla-Jensen, S.,** Die Hauptlinien des naturlichen Bacteriensystem, *Zentralbl. Bakteriol. Parasitend. Infektionskr.*, Abt. 1, 22, 305, 1909.
13. **Dworkin, M. and Foster, J. W.,** Studies on *Pseudomonas methanica* (Sohngen) nov. comb., *J. Bacteriol.*, 72, 592, 1956.
14. **Foster, J. W. and Davis, R. H.,** A methane-dependent coccus, with notes on classification of obligate, methane-utilizing bacteria, *J. Bacteriol.*, 91, 1924, 1966.
15. **Davis, J. B., Coty, V. F., and Stanley, J. P.,** Atmospheric nitrogen fixation by methane-utilizing bacteria, *J. Bacteriol.*, 88, 468, 1964.
16. **Brown, L. R., Stravinski, R. J., and McCleskey, C. S.,** The isolation and characterizaion of *Methanomonas methanooxidans, Can.J. Microbiol.*, 10, 791, 1964.
17. **Stocks, P. K. and McCleskey, C. S.,** Morphology and physiology of *Methanomonas methanooxidans, J. Bacteriol.*, 88, 1071, 1964.
18. **Proctor, H. M., Harris, J. R., and Ribbons, D. W.,** Fine structure of methane-utilizing bacteria, *J. Appl. Bacteriol.*, 32, 118, 1969.
19. **Smith, J. and Ribbons, D. W.,** Fine structure of *Methanomonas methanoooxidans, Arch. Mikrobiol.*, 74, 116, 1970.
20. **Davies, S. L. and Whittenbury, R.,** Fine structure of methane and other hydrocarbon utilising bacteria, *J. Gen. Microbiol.*, 61, 227, 1970.
21. **Romanovskaya, V. A., Sadovmikov, Y. S., and Malashenko, Y. R.,** Formation of taxons of methane-oxidizing bacteria using numerical analysis methods, *Microbiology (USSR)*, 47, 120, 1978.
22. **Patt, T. E., Cole, G. C., Bland, J., and Hanson, R. S.,** Isolation and characterization of bacteria that grow on methane and organic compounds as sole sources of carbon and energy, *J. Bacteriol.*, 120, 955, 1974.
23. **Patel, R. N., Hou, C. T., and Felix, A.,** Microbial oxidation of methane and methanol: isolation of methane-utilizing bacteria and characterization of a facultative methane-utilizing isolate, *J. Bacteriol.*, 136, 352, 1978.
24. **Romanovskaya, V. A., Malashenko, Yu. R., and Bogachenko, V. N.,** Corrected diagnosis of the genera and species of methane-utilizing bacteria, *Mikrobiologiya*, 47, 96, 1978.
25. **Whittenbury, R., Colby, J., Dalton, H., and Reed, H. L.,** *Microbial Production and Utilization of Gases* Schlegel, H. G., Gottschalk, G., and Pfennig, N., Eds., Erich Goltze, Gottingen, 1976, 281.
26. **Trotsenko, Yu. A.,** *Microbial Production and Utilization of Gases* Schlegel, G., Gottschalk, G., and Pfennig, N. Eds., Erich Goltze, Gottingen, 329, 1975.
27. **Taylor, S. C., Dalton, H., and Down, C. S.,** Ribulose 1,5-bisphosphate carboxylase/oxygenase and carbon assimilation in *Methylococcus capsulatus* (Bath), *J. Gen. Microbiol.*, 122, 89, 1981.
28. **Bamford, C. W. and Quayle, J. R.,** Hydroxypyruvate reductase activity in *Paracoccus denitrificans, J. Gen. Microbiol.*, 101, 259, 1977.
29. **Hou, C. T., Patel, R. N., Laskin, A. I., Barnabe, N., and Marczak, I.,** Microbial oxidation of gaseous hydrocarbons: production of methyl ketones from their corresponding secondary alcohols by methane- and methanol-grown microbes, *Appl. Environ. Microbiol.*, 38, 135, 1979.

30. **Dalton, H. and Whittenbury, R.,** Nitrogen metabolism in *Methylococcus capsulatus* strain Bath, in *Symp. Microb. Prod. Utilisation Gases (H_2, CH_4, CO)*, Schlegel, H. G., Gottschalk, G. and Pfenning, N., Eds., Akademie der Wissenschafter, Gottingen, 1965, 379.
31. **deBont, J. A. M. and Mulder, E. G.,** Nitrogen fixation and co-oxidation of ethylene by a methane-utilising bacterium, *J. Gen. Microbiol.*, 83, 113, 1974.
32. **Shishkina, V. N. and Trotsenko, Y. A.,** Pathways of ammonia assimilation in obligate methane utilizers, *FEMS Microbiol. Lett.*, 5, 187, 1979.
33. **Whittenbury, R., Davies, S. L., and Davey, J. G.,** Exospores and cysts formed by methane-utilizing bacteria, *J. Gen. Microbiol.*, 61, 219, 1970.
34. **DeBoer, W. E. and Hazeu, W.,** Observations on the fine structure of a methane-utilizing bacterium, *Antonie van Leeuwenhoek, J. Microbiol. Seriol.*, 38, 33, 1972.
35. **Hazeu, W. and Batenburg-van der Egte, W. M.,** Some characteristics of *Methylococcus mobilis* sp. nov., *Arch. Microbiol.*, 124; 211, 1980.
36. **Hyder, S. L., Mayers, A., and Cayer, M. L.,** Membrane modulation in a methylotrophic bacterium *Methylococcus capsulatus* (Texas) as a function of growth substrate, *Tissue Cell*, 11, 597, 1979.
37. **Smith, U., Ribbons, D. W., and Smith, D. S.,** The fine structure of *Methylococcus capsulatus*, *Tissue Cell*, 2, 513, 1970.
38. **Galachenko, V. F. and Suzina, N. E.,** Peculiarities of membrane apparatus organisation of methanotrophic bacteria, in *Microbial Growth on C_1-Compounds*, Skryabin, G. K., Ivanov, M. V., Kondratjeva, E. N., Zavarzin, G. A., Trotsenko, Yu. A., and Nesterov, A. I., Eds., Abst. 2nd Intl. Symposium on Microbial Growth on C_1-compounds, U.S.S.R. Academy of Sciences, Pushchino, U.S.S.R., 1977, 12.
39. **Linton, J. D. and Vokes, J.,** Growth of the methane-utilizing baterium *Methylococcus* NCIB in mineral salts medium with methanol as the sole source of carbon, *FEMS Microbiol. Lett.*, 4, 125, 1978.
40. **Monosov, E. Z.,** Membrane genesis in methane oxidizing bacteria, in *Microbial Growth on C_1-Compounds*, Skryabin, G. K., Ivanov, M. V., Kondratjeva, E. N., Zavarzin, G. A., Trotsenko, Yu. A., and Nesterov, A. I. Eds., Abst. 2nd Intl. Symposium on Microbial Growth on C_1-compounds, U.S.S.R. Academy of Sciences, Pushchino, U.S.S.R. 1977, 25.
41. **Weaver, T. L. and Dugan, P. R.,** Ultrastructure of *Methylosinus trichosporium* as revealed by freeze etching, *J. Bacteriol.*, 121, 704, 1975.
42. **Best, D. J. and Higgins, I. J.,** Methane-oxidizing activity and membrane morphology in a methanol-grown obligate methanotroph, *Methylosinus trichosporium* OB3b, *J. Gen. Microbiol.*, 125, 73, 1981.
43. **Brannan, J. and Higgins, I. J.,** Effect of growth conditions on the intracytoplasmic membranes of *Methylosinus trichosporium* OB3b, *Proc. Soc. Gen. Microbiol.*, 5, 69, 1978.
44. **Scott, D., Brannan, J., and Higgins, I. J.,** The effect of growth conditions on intracytoplasmic membranes and methane mono-oxygenase activities in *Methylosinus trichosporium* OB3b, *J. Gen. Microbiol.*, 125, 63, 1981.
45. **Suzina, N. E. and Fikhte, B. A.,** Fine structure of tubular formations found out on the surface of cells *Methylocystis echinoides* sp. nov., in *Microbial Growth on C_1-Compounds*, Skryabin, G. K., Ivanov, M. V., Kondratjeva, E. N., Zavarzin, G. A., Trotsenio, Yu. A., and Nesterov, A. I. Eds., Abst. 2nd Intl. Symposium on Microbial Growth on C_1-compounds, U.S.S.R. Academy of Sciences, Pushchino, U.S.S.R., 1977, 23.
46. **Haubold, R.,** Two different types of surface structures of methane-utilizing bacteria, *Z. Allg. Mikrobiol.*, 18, 511, 1978.
47. **Patt, T. E., Cole, G. C., and Hanson, R. S.,** *Methylobacterium*, a new genus of facultatively methylotrophic bacteria, *Int. J. Syst. Bacteriol.*, 26, 226, 1976.
48. **Patt, T. E. and Hanson, R. S.,** Intracytoplasmic membrane, phospholipid, and sterol content of *Methylobacterium organophilum* cells grown under different conditions, *J. Bacteriol.*, 134, 636, 1978.
49. **Ferenci, T.,** Carbon monoxide-stimulated respiration in methane-utilizing bacteria, FEBS Lett., 41, 94, 1974.
50. **Ferenci, T., Strom, T., and Quayle, J. R.,** Oxidation of carbon monoxide and methane by *Pseudomonas methanica*, *J. Gen. Microbiol.*, 91, 79, 1975.
51. **Ribbons, D. W.,** Oxidation of C_1 compounds by particulate fractions from *Methylococcus capsulatus*: distribution and properties of methane-dependent reduced nicotinamide adenine dinucleotide oxidase (methane hydroxylase), *J. Bacteriol.*, 122, 1351, 1975.
52. **Ribbons, D. W. and Michalover, J. L.,** Methane oxidation of cell free extracts of *Methylococcus capsulatus*, *FEBS Lett.*, 11, 41, 1970.
53. **Tonge, G. M., Harrison, D. E. F., Knowles, G. J., and Higgins, I. J.,** Properties and partial purification of the methane-oxidizing enzyme system from *Methylosinus trichosporium*, *FEBS Lett.*, 59, 293, 1975.
54. **Wadzinksi, A. M. and Ribbons, D. W.,** Oxidation of C_1 compounds by particulate fractions from *Methylococcus capsulatus*: properties of methanol oxidase and methanol dehydrogenase, *J. Bacteriol.*, 122, 1364, 1975.

55. **Patel, R. N., Hou, C. T., Laskin, A. I., Felix, A., and Derelanko, P.,** Microbial oxidation of gaseous hydrocarbons. II. Hydroxylation of alkanes and epoxidation of alkenes by cell-free particulate fractions of methane-utilizing bacteria, *J. Bacteriol.*, 139, 675, 1979.
56. **Patel, R. N., Hou, C. T., Laskin, A. I., and Felix, A.,** Microbial oxidation of hydrocarbons: properties of a soluble methane monooxygenase from a facultative methane-utilizing organism, *Methylobacterium* sp. CRL-26, *Appl. Environ. Microbiol.*, 44, 1130, 1982.
57. **Takeda, K. and Tanaka, K.,** Ultrastructure of intracytoplasmic membranes of *Methanomonas margaritae* cells grown under different conditions, *Antonie van Leeuwenhoek, J. Microbiol. Serol.*, 46, 15, 1980.
58. **Lawrence, A. J. and Quayle, J. R.,** Alternative carbon assimilation pathways in methane-utilizing bacteria, *J. Gen. Microbiol.*, 63, 371, 1970.
59. **Lynch, M. J., Wopat, A. E., and O'Connor, M. L.,** Characterization of two new facultative methanotrophs, *Appl. Environ. Microbiol.*, 40, 400, 1980.
60. **Green, P. N. and Bousfield, I. J.,** A taxonomic study of some Gram-negative facultatively methylotrophic bacteria, *J. Gen. Microbiol.*, 128, 623, 1982.
61. **O'Connor, M. L.,** Extension of the model concerning linkage of genes coding for C_1-related functions in *Methylobacterium organophilum*, *Appl. Environ. Microbiol.*, 41, 437, 1981.
62. **Makula, R. A.,** Phospholipid composition of methane-utilizing bacteria, *J. Bacteriol.*, 134, 771, 1978.
63. **Weaver, T. L., Patrick, M. A., and Dugan, P. R.,** Whole cell and membrane lipids of the methylotrophic bacterium *Methylosinus trichosporium*, *J. Bacteriol.*, 124, 602, 1975.
64. **Bird, C. W., Lynch, J. M., Pirt, S. J., Reid, W. W., Brooks, C. J. W., and Middleditch, B. S.,** Sterols and squalene in *Methylosinus capsulatus* grown on methane, *Nature (London)*, 230, 473, 1971.
65. **Hagen, P. O., Goldfine, H., and Williams, P. J. L.,** Phospholipids of bacteria with extensive intracytoplasmic membranes, *Science*, 151, 1543, 1966.
66. **Davey, J. F., Whittenbury, R., and Wilkinson, J. F.,** The distribution in the methylotrophic bacteria of some key enzymes concerned with intermediary metabolism, *Arch. Microbiol.*, 87, 359, 1972.
67. **Patel, R. N., Hoare, S. L., Hoare, D. S., and Taylor, B. F.,** Incomplete tricarboxylic acid cycle in a type I methylotroph, *Methylococcus capsulatus*, *J. Bacteriol.*, 123, 382, 1975.
68. **Lawrence, A. J., Kemp, M. B., and Quayle, J. R.,** Synthesis of cell constituents by methane-grown *Methylococcus capsulatus* and *Methanomonas methanooxidans*, *Biochem. J.*, 116, 631, 1970.
69. **Kato, N., Tsuji, K., Ohashi, H., Tani, Y., and Ogata, K.,** The assimilation pathway of C_1-compounds in *Streptomyces* sp. No. 239 during growth on methanol, *Agric. Biol. Chem.*, 41, 29, 1977.
70. **Shishkina, V. N., Yurchenko, V. V., Romanovskaya, V. A., Malashenko, Yu. R., and Trotsenko, Yu. A.,** Alternate pathways of methane assimilation in obligate methanotrophs, *Microbiology*, 45, 359, 1976.
71. **Strom, T., Ferenci, T., and Quayle, J. R.,** The carbon assimilation pathways of *Methylococcus capsulatus*, *Pseudomonas methanica*, and *Methylosinus trichosporium* (OB3b) during growth on methane, *Biochem J.*, 144, 465, 1974.
72. **Salem, A. R., Hacking, A. R., and Quayle, J. R.,** Cleavage of malylcoenzyme A into acetyl coenzyme A and glyoxylate by *Pseudomonas* AMI and other C_1-unit-oxidising bacteria, *Biochem. J.*, 136, 89, 1973.
73. **Whittenbury, R. and Kelly, D. P.,** In Microbial Energetics, Symp. 27. The Soc. Gen. Microbiol., Haddock, B. A., Hamilton, W. A., Eds., pp. 121, Cambridge: Cambridge Univ. Press.
74. **Quayle, J. R. and Ferenci, T.,** Evolutionary aspects of autotrophy, *Microbiol. Rev.*, 42, 251, 1978.
75. **Leadbetter, E. R. and Foster, J. W.,** Studies on some methane-utilizing bacteria, *Arch. Mikrobiol.*, 30, 91, 1958.
76. **Johnson, P. A. and Quayle, J. R.,** Microbial growth on C_1 compounds. Synthesis of constituents by methane- and methanol-grown *Pseudomonas methanica*, *Biochem. J.*, 95, 859, 1965.
77. **Kemp, M. B. and Quayle, J. R.,** "Microbial growth on C_1 compounds. Uptake of ^{14}C-formaldehyde and ^{14}C-formate by methane-grown *Pseudomonas methanica* and determination of the hexose labeling pattern after incubation with ^{14}C-methanol, *Biochem. J.*, 102, 94, 1967.
78. **Kemp, M. B. and Quayle, J. R.,** Incorporation of C_1 units into allulose phosphate by methane-grown *Pseudomonas methanica*, *Biochim. Biophys. Acta*, 107, 174, 1965.
79. **Kemp, M. B. and Quayle, J. R.,** Microbial growth on C_1 compounds. Incorporation of C_1 units into allulose phosphate by extracts of *Pseudomonas methanica*, *Biochem. J.*, 99, 41, 1966.
80. **Kemp, M. B.,** Hexose phosphate synthase from *Methylococcus capsulatus* makes D-arabino-3-hexulose phosphate, *Biochem. J.*, 139, 129, 1974.
81. **Ferenci, T., Strom, T., and Quayle, J. R.,** Purification and properties of 3-hexulose phosphate synthase and phospho-3-hexuloisomerase from *Methylococcus capsulatus*, *Biochem. J.*, 144, 477, 1974.
82. **Sahm, H., Schutte, H., and Kula, M. R.,** Purification and properties of 3-hexulose phosphate synthase from *Methylomonas* M15, *Eur. J. Biochem.*, 66, 591, 1976.
83. **Kato, N., Ohashi, H., Tani, Y., and Ogata, K.,** 3-Hexulose phosphate synthase from *Methylomonas aminofaciens* 77a. Purification, properties and kinetics, *Biochim. Biophys. Acta*, 523, 236, 1978.

84. **Colby, J. and Zatman, L. J.,** Enzymological aspects of the pathways of trimethylamine oxidation and C_1 assimilation in obligate methylotrophs and restricted facultative methylotrophs, *Biochem. J.*, 148, 513, 1975.
85. **Loginova, N. V. and Trotsenko, Yu. A.,** Microbial Growth on C_1-Compounds, Skryabin, G. K., Ivanov, M. V., Kondratjeva, E. N., Zavarzin, G. A., Trotsenko, Yu. A., and Nesterov, A. I., Eds., U.S.S.R. Academy of Sciences, Puschino, 1977, 37.
86. **Babel, W. and Miethe, D.,** Alternative reaction sequence of the cellulose-6-phosphate pathway for a methylotrophic bacterium, *Z. Alleg. Mikrobiol.*, 14, 153, 1974.
87. **Ben Bassat, A. and Goldberg, I.,** Oxidation of C_1 compounds in *Pseudomonas* C, *Biochim. Biophys. Acta*, 497, 586, 1977.
88. **Beardsmore, A. J., Aperghis, P. N. G., and Quayle, J. R.,** Characterization of the assimilatory and dissimilatory pathways of carbon metabolism during growth of *Methylophilus methylotrophus* on methanol, *J. Gen. Microbiol.*, 128, 1423, 1982.
89. **Ben-Bassat, A., Goldberg, I., and Mateles, R. I.,** Distribution of methanol carbon between assimilation and oxidation pathways in methanol-grown *Pseudomonas* C., *J. Gen. Microbiol.*, 116, 213, 1980.
90. **Trotsenko, Y. A.,** "Isolation and characterisation of obligate methanotrophic bacteria" p. 329, 1976. In Schlegel, H. G., Gottschalk, G., and Pfennig, N., Eds., "Symposium on microbial production and utilization of gases (H_2, CH_4, CO), Gottingen. Akademie der Wissenschaften, Gottingen.
91. **Large, P. J., Peel, D., and Quayle, J. R.,** Microbial growth on C_1 compounds. II. Synthesis of cell constituents by methanol- and formate-grown *Pseudomonas* AM1 and methanol-grown *Hyphomicrobium vulgare*, *Biochem. J.*, 81, 470, 1961.
92. **Large, P. J,., Peel, D., and Quyale, J. R.,** Microbial growth on C_1 compounds. III. Distribution of radioactivity in metabolites of methanol-grown *Pseudomonas* AM1 after incubation with [^{14}C] methanol and [^{14}C] bicarbonate, *Biochem. J.*, 82, 483, 1962.
93. **Weaver, T. L. and Dugan, P. R.,** Methylotrophic enzyme distribution in *Methylosinus trichosporium*, *J. Bacteriol.*, 122, 433, 1975.
94. **O'Connor, M. S. and Hanson, R. S.,** Serine transhydroxymethylase isoenzymes from a facultative methylotroph, *J. Bacteriol.*, 124, 985, 1975.
95. **Blackmore, M. A. and Quayle, J. R.,** Microbial growth on oxalate by a route not involving glyoxylate carboligase, *Biochem. J.*, 118, 53, 1970.
96. **Large, P. J. and Quayle, J. R.,** Microbial growth on C_1 compounds. V. Enzyme activities in extracts of *Pseudomonas* AM1, *Biochem. J.*, 87, 386, 1963.
97. **Salem, A. R., Large, P. J., and Quayle, J. R.,** Glycine formation during growth of *Pseudomonas* AM1 on methanol and succinate, *Biochem. J.*, 128, 1203, 1972.
98. **Heptinstall, J. and Quayle, J. R.,** Pathway leading to and from serine during growth of *Pseudomonas* AM1 on C_2 compounds or succinate, *Biochem. J.*, 117, 563, 1970.
99. **Harder, W., Attwood, M. M., and Quayle, J. R.,** Methanol assimilation by *Hyphomicrobium* spp., *J. Gen. Microbiol.*, 78, 155, 1973.
100. **Large, P. J., Peel, D., and Quayle, J. R.,** Microbial growth on C_1 compounds. IV. Carboxylation of phosphoenolypyruvate in methanol-grown *Pseudomonas* AM1, *Biochem. J.*, 85, 243, 1962.
101. **Loginova, N. V. and Trotsenko, Y. A.,** Pyruvate and phosphoenolpyruvate carboxylases in methylotrophs, *Microbiology (U.S.S.R.)* 48, 202, 1979.
102. **Naguib, M.,** Alternative carboxylation reactions in type II methylotrophs and the localisation of carboxylase activities in intracytoplasmic membranes, *Z. Allg. Mikrobiol.*, 19, 333, 1979.
103. **Hersh, L. B. and Bellion, E.,** Malate cleavage reaction in *Pseudomonas* species (Shaw strain MA), *Biochem. Biophys. Res. Commun.*, 48, 712, 1972.
104. **Cox, R. B. and Zatman, L. J.,** Isocitrate lyase and adenosine triphosphate malate lyase as key enzymes for the methylotrophic growth of bacterium $5H_2$, *Biochem. Soc. Trans.*, 1, 669, 1973.
105. **Hersh, L. B.,** Malate adenosine triphosphate lyase. Separation of the reaction into a malate thiokinase and malyl coenzyme A lyase, *J. Biol. Chem.*, 248, 7295, 1973.
106. **Dunstan, P. M., Anthony, C., and Drabble, W,.** Microbial metabolism of C_1 and C_2 compounds. The role of glyoxylate, glyocollate and acetate in the growth of *Pseudomonas* AM1 on ethanol and on C_1 compounds, *Biochem. J.*, 128, 107, 1972.
107. **Bellion, E. and Hersh, L. B.,** Methylamine metabolism in a *Pseudomonas* species, *Arch. Biochem. Biophys.*, 153, 368, 1972.
108. **Large, P. J. and Carter, R. H.,** Specific activities of enzymes of the serine pathway of carbon assimilation in *Pseudomonas aminovorans* and *Pseudomonas* MS grown on methylamine, *Biochem. Soc. Trans.*, 1, 1973.
109. **Quayle, J. R.,** Unsolved problems in the microbial metabolism of methane and methanol, in *Microbial Growth on C_1 Compounds*, Terui, G., Ed., Society of Fermentation Technology, Tokyo, Japan, 1975, 59.

110. **Bellion, E.,** Investigations into the ICL serine pathway, in *Microbial Growth on C_1-Compounds*, Skryabin, G. K., Ivanov, M. V., Kondratjeva, E. N., Zavarzin, G. A., Trotsenko, Yu. A., and Nesterov, A. I. Eds., Abst. 2nd Intl. Symposium on Microbial Growth on C_1-Compounds, U.S.S.R. Academy of Sciences, Pushchino, U.S.S.R., 1977, 36.
111. **Kortstee, G. J. J.,** The homoisocitrate-glyoxylate cycle in pink, facultative methylotrophs, *FEMS Microbiol. Lett.*, 8, 59, 1980.
112. **Harrington, A. A. and Kallio, R. E.,** Oxidation of methanol and formaldehyde by *Pseudomonas methanica, Can. J. Microbiol.*, 6, 1, 1960.
113. **Kallio, R. E. and Harrington, A. A.,** Sudanophilic granules and lipids of *Pseudomonas methanica, J. Bacteriol.*, 80, 321, 1960.
114. **Hazeu, W., and Steenis, P. J.,** Isolation and characterization of two vibrio shaped methane oxidizing bacteria, *Antoine van Leeuwenhoek. J. Microbiol. Serol.*, 36, 67, 1970.
115. **Malashenko, Y. R.,** Isolation and characterization of new species (thermophilic and thermotolerant ones) of methane-utilizers. p. 293, In Schlegel, H. G., Gottschalk, G., and Pfennig, N. Ed., Symposium on microbial production and utilization in gases (H_2, CH_4, CO). Gottingen, Akademie der Wissenschaften, Gottingen.
116. **Thompson, A. W., O'Neil, J. G. O., and Wilkinson, J. F.,** Acetone production by methylobacteria, *Arch. Mikrobiol.*, 109, 243, 1976.
117. Imperial Chem. Inds. Ltd., Preparation of poly-3-hydroxybutyric acid by aerobic cultivation of purified *Methylobacterium organophilum*, British Patent 006,079, 1979.
118. **Leadbetter, E. R. L. and Foster, J. W.,** Incorporation of molecular oxygen in bacterial cells utilizing hydrocarbons for growth, *Nature (London)*, 184, 1428, 1959.
119. **Higgins, I. J. and Quayle, J. R.,** Oxygenation of methane by methane-grown *Pseudomonas methanica* and *Methanomonas methanooxidans, Biochem. J.*, 118, 201, 1970.
120. **Hubley, J. H., Thomson, A. W., and Wilkinson, J. F.,** Specific inhibitors of methane oxidation in *Methylosinus trichosporium, Arch. Microbiol.*, 102, 199, 1975.
121. **Patel, R. N., Hou, C. T., and Felix, A.,** Inhibition of dimethyl ether and methane oxidation in *Methylococcus capsulatus* and *Methylosinus trichosporium, J. Bacteriol.*, 126, 1017, 1976.
122. **Colby, J., Dalton, H., and Whittenbury, R.,** An improved assay for bacterial methane monooxygenase: some properties of the enzyme from *Methylomonas methanica, Biochem. J.*, 151, 459, 1975.
123. **Tonge, G. M., Knowles, C. J., Harrison, D. E. F., and Higgins, I. J.,** Metabolism of one carbon compounds. Cytochromes of methane and methanol-utilizing bacteria, *FEBS Lett.*, 44, 106, 1975.
124. **Tonge, G. M., Harrison, D. E. F., Knowles, C. J., and Higgins, I. J.,** Properties and partial purification of the methane-oxidising enzyme system from *Methylosinus trichosporium, FEBS Lett.*, 58, 293, 1975.
125. **Tonge, G. M., Harrison, D. E. F., and Higgins, I. J.,** Purification and properties of the methane monooxygenase enzyme system from *Methylosinus trichosporium* OB3b, *Biochem. J.*, 161, 333, 1977.
126. **Hou, C. T., Patel, R. N., Laskin, A. I., and Barnabe, N.,** Microbial oxidation of gaseous hydrocarbons: epoxidation of C_2 to C_4 *n*-alkenes by methylotrophic bacteria, *Appl. Environ. Microbiol.*, 38(1), 127, 1979.
127. **Patel, R. N., Hou, C. T., Laskin, A. I., Felix, A., and Derelanko, P.,** Microbial oxidation of gaseous hydrocarbons: production of methylketones from corresponding *n*-alkanes by methane-utilizing bacteria, *Appl. Environ. Microbiol.*, 39, 727, 1980.
128. **Patel, R. N., Hou, C. T., Laskin, A. I., Felix, A., and Derelanko, P.,** Microbial oxidation of gaseous hydrocarbons: production of secondary alcohols from corresponding *n*-alkanes by methane-utilizing bacteria, *Appl. Environ. Microbiol.*, 39, 720, 1980.
129. **Colby, J. and Dalton, H.,** Some properties of a soluble methane monooxygenase from *Methylococcus capsulatus* strain Bath, *Biochem. J.*, 157, 495, 1976.
130. **Colby, J., Stirling, D. I., and Dalton, H.,** The soluble methane monooxygenase of *Methylococcus capsulatus* (Bath). Its ability to oxygenate *n*-alkanes, *n*-alkenes, ethers, alicyclic, aromatic, and heterocyclic compounds, *Biochem. J.*, 165, 395, 1977.
131. **Colby, J. and Dalton, H.,** Resolution of the methane monooxygenase of *Methylococcus capsulatus* (Bath) into three components. Purification and properties of component C, flavoprotein, *Biochem. J.*, 171, 461, 1978.
132. **Stirling, D. I. and Dalton, H.,** Properties of the methane monooxygenase from extracts of *Methylosinus trichosporium* (OB3b) and evidence for its similarity to the enzyme from *Methylosinus capsulatus* (Bath), *Eur. J. Biochem.*, 96, 205, 1979.
133. **Colby, J. and Dalton, H.,** Characterization of the second prosthetic group of the flavoenzyme NADH-acceptor (component C) of the methane monooxygenase from *Methylococcus capsulatus* (Bath), *Biochem. J.*, 177, 903, 1979.
134. **Hutchinson, D. W., Whittenbury, R., and Dalton, H.,** A possible role of free radicals in the oxidation of methane by *Methylococcus capsulatus, J. Theor. Biol.*, 58, 325, 1976.

135. **Dalton, H.,** Methane monooxygenase from a variety of microbes, p 1, In Dalton, H., (Ed.) Microbial growth on C_1 compounds Proc. of the third international symposium on microbial growth on C_1 compounds, Sheiffeld, United Kingdom, Hayden & Son, London.
136. **Anthony, C. and Zatman, L. J.,** The microbial oxidation of methanol. I. Isolation and properties of *Pseudomonas* M27, *Biochem. J.*, 92, 609, 1964.
137. **Anthony, C. and Zatman, L. J.,** The microbial oxidation of methanol. II. The methanol-oxidising enzyme of *Pseudomonas* sp. M27, *Biochem. J.*, 92, 614, 1964.
138. **Anthony, C. and Zatman, L. J.,** The microbial oxidation of methanol. The methanol oxidizing enzyme of *Pseudomonas* sp. M27, *Biochem. J.*, 96, 808, 1964.
139. **Anthony, C. and Zatman, L. J.,** The microbial oxidation of methanol. Purification and properties of the alcohol dehydrogenase of *Pseudomonas* M27, *Biochem. J.*, 1-4, 953, 1967.
140. **Patel, R. N., Bose, H. R., Mandy, W. J., and Hoare, D. S.,** Physiological studies of methane- and methanol-oxidizing bacteria: comparison of a primary alcohol dehydrogenase from *Methylococcus capsulatus* (Texas) and *Pseudomonas* M27, *J. Bacteriol.*, 110, 570, 1972.
141. **Sperl, G. T., Forrest, H. S., and Gibson, D. T.,** Substrate specificity of the purified primary alcohol dehydrogenases from methanol-oxidizing bacteria, *J. Bacteriol.*, 118, 541, 1974.
142. **Yamanaka, K. and Matsumoto, T.,** Purification, crystallization and properties of primary alcohol dehydrogenase from a methanol-oxidizing *Pseudomonas* p. No. 2941, *Agric. Biol. Chem.*, 41, 467, 1977.
143. **Duine, J. A., Frank, J., and Westerling, J.,** Purification and properties of methanol dehydrogenase from *Hyphomicrobium* X, *Biochim. Biophys. Acta*, 524, 277, 1978.
144. **Yamanaka, K., and Matsumoto, T.,** In *Microbial Growth on C_1-compounds*, Skryabin, G. K., Ivanov, M. V., Kondratjeva, E. N., Zavarzin, G. A., Trotsenko, Yu. A., and Neterov, A. I., Eds., U.S.S.R. Academy of Sciences, Puschino, 1977, 72.
145. **Metha, R. J.,** Studies on methanol-oxidizing bacteria. II. Purification and properties of methanol dehydrogenase from *Pseudomonas* RJ1, *Antonie van leeuwenhoek, J. Microbiol. Serol.*, 39, 303, 1973.
146. **Cox, R. B. and Quayle, J. R.,** The autotrophic growth of *Micrococcus denitrificans* on methanol, *Biochem. J.*, 150, 569, 1975.
147. **Goldberg, I.,** Purification and properties of a methanol oxidizing enzyme in *Pseudomonas* C, *Eur. J. Biochem.*, 63, 233, 1976.
148. **Wolf, H. J. and Hanson, R. S.,** Alcohol dehydrogenase from *Methylobacterium organophilum, Appl. Environ. Microbiol.*, 36, 105, 1978.
149. **Patel, R. N. and Felix, A.,** Microbial oxidation of methane and methanol: crystallization and properties of methanol dehydrogenase from *Methylosinus sporium, J. Bacteriol.*, 128, 413, 1976.
150. **Patel, R. N., Hou, C. T., and Felix, A.,** Microbial oxidation of methane and methanol: crystallization of methanol dehydrogenase and properties of holo- and apo-methanol dehydrogenase from *Methylomonas methanica, J. Bacteriol.*, 133, 641, 1978.
151. **Bamforth, C. W. and Quayle, J. R.,** The dye-linked alcohol dehydrogenase of *Rhodopseudomonas acidophila*. Comparison with dye-linked methanol dehydrogenase, *Biochem. J.*, 169, 677, 1978.
152. **Duine, J. A., Frank, J., and Verwiel, P. E. J.,** Isolation of methanol dehydrogenase with a functional coupling to cytochrome C, *J. Gen. Microbiol.*, 115, 523, 1979.
153. **Anthony, C. and Zatman, L. J.,** The microbial oxidation of methanol. The prosthetic group of the alcohol dehydrogenase of *Pseudomonas* M27: a new oxidoreductase prosthetic group, *Biochem. J.*, 104, 960, 1967.
154. **DeBeer, R., van Ormondt, D., Van Ast, M. A., Baneu, R., Duine, J. A., and Frank, J.,** ESR and ENDOR at 9 and 35 GH_2 on a powder of the enzyme methanol dehydrogenase from *Hyphomicrobium* X, *J. Chem. Phys.*, 70, 4491, 1979.
155. **Westerling, J., Frank, J., and Duine, J. A.,** The prosthetic group of methanol dehydrogenase from *Hyphomicrobium* X: electron spin resonance evidence for a quinone structure, *Biochim. Biophy. Acta*, 87, 719, 1979.
156. **Salisbury, S. A., Forrest, H. S., Cruse, W. B. T., and Kennard, O.,** A novel coenzyme from bacterial dehydrogenases, *Nature (London)*, 280, 843, 1979.
157. **Duine, J. A., Frank, J., and Verwiel, P. E. J.,** Structure and activity of the prosthetic group of methanol dehydrogenase, *Eur. J. Biochem.*, 108, 187, 1980.
158. **Forrest, H. S., Salisbury, S. A., and Kilty, C. G.,** A mechanism for the enzymic oxidation of methanol involving methoxatin, *Biochem. Biophys. Res. Commun.*, 97, 248, 1980.
159. **Duine, J. A., Frank, J., and Verwiel, P. E. J.,** Characterization of the second prosthetic group in methanol dehydrogenase from *Hyphomicrobium* X, *Eur. J. Biochem.*, 118, 395, 1981.
160. **Mincey, Y., Bell, J. A., Mildvan, A. S., and Abeles, R. H.,** Mechanism of action of methoxatin-dependent alcohol dehydrogenase, *Biochemistry*, 20, 7502, 1981.
161. **Johnson, P. A. and Quayle, J. R.,** Microbial growth on C_1 compounds. VI. Oxidation of methanol, formaldehyde, and formate by methanol-grown *Pseudomonas* AM1, *Biochem. J.*, 93, 281, 1964.
162. **Hampton, D. and Zatman, L. J.,** The metabolism of tetramethylammonium chloride by Bacterium 5H2, *Biochem. Soc. Trans.*, 1, 667, 1973.

163. **Boulton, C. A. and Large, P. J.,** Synthesis of certain assimilatory and dissimilatory enzymes during bacterial adaptation to growth on trimethylamine, *J. Gen. Microbiol.*, 101, 151, 1977.
164. **Kung, H. F. and Wagner, C.,** Oxidation of C_1 compounds by *Pseudomonas* sp. Ms., *Biochem. J.*, 116, 357, 1970.
165. **O'Connor, J. L. and Hanson, R. S.,** Enzyme regulation in *Methylobacterium organophilum, J. Gen. Microbiol.*, 101, 327, 1977.
166. **Colby, J. and Zatman, L. J.,** Trimethylamine metabolism in obligate and facultative methylotrophs, *Bioch. J.*, 132, 101, 1973.
167. **Stirling, D. I. and Dalton, H.,** Purification and properties of an NAD(P)$^+$-linked formaldehyde dehydrogenase from *Methylococcus capsulatus* (Bath), *J. Gen. Microbiol.*, 107, 19, 1978.
168. **Metha, R. J.,** A novel inducible formaldehyde dehydrogenase of *Pseudomonas* sp. (RJ_1), *Antonie van Leeuwenhoek*, 41, 89, 1975.
169. **Patel, R. N. and Hoare, D. S.,** Physiological studies of methane and methanol-oxidizing bacteria: oxidation of C_1 compounds by *Methylococcus capsulatus, J. Bacteriol.*, 107, 187, 1971.
170. **Patel, R. N., Hoare, S. L., and Hoare, D. S.,** [1-^{14} C]-Acetate assimilation by obligate methanotrophs, *Pseudomonas methanica* and *Methylosinus trichosporium, Antonie van Leeuwenhoek*, 45, 499, 1979.
171. **Patel, R. N., Hou, C. T., and Felix, A.,** Microbial oxidation of methane and methanol: purification and properties of a heme-containing aldehyde dehydrogenase from *Methylomonas methylovora, Arch. Microbiol.*, 122, 241, 1979.
172. **Patel, R. N., Hou, C. T., Derelanko, P., and Felix, A.,** Purification and properties of a heme-containing aldehyde dehydrogenase from *Methylosinus trichosporium, Arch. Biochem. Biophys.*, 203, 654, 1980.
173. **Marison, I. W. and Attwood, M. M.,** Partial purification and characterization of a dye-linked formaldehyde dehydrogeanse from Hyphomicrobium X, *J. Gen. Microbiol.*, 117, 305, 1980.
174. **Attwood, M. M. and Harder, W.,** Formate assimilation of *Hyphomicrobium* X, *FEMS Microbiol. Lett.*, 3, 111, 1978.
175. **Rabinowitz, J. C. and Pricer, W. E.,** Formyl tetrahydrofolate synthetase, *Methods in Enzymology*, 6, 375, 1963.
176. **Whiteley, H. R. and Huennekens, F. M.,** Mechanism of the reaction catalyzed by the formate-activating enzyme from *Micrococcus aerogenes, J. Biol. Chem.*, 237, 1290, 1962.
177. **O'Brien, W. E., Brewer, J. M., and Ljungdahl, L. G.,** Purification and characterization of thermostable 5,10-methylenetetrahydrofolate dehydrogenase from *Clostridium thermoaceticum, J. Biol. Chem.*, 248, 403, 1973.
178. **Marison, I. W. and Attwood, M. M.,** A possible alternative mechanism for the oxidation of formaldehyde to formate, *J. Gen. Microbiol.*, 128, 1441, 1982.
179. **Andreesen, J. R. and Ljungdahl, L. G.,** Nicotinamide adenine dinucleotide phosphate-dependent formate dehydrogenase from *Clostridium thermoaceticum*: purification and properties, *J. Bacteriol.*, 120, 6, 1974.
180. **Scherer, P. A. and Thauer, R. K.,** Purification and properties of reduced ferredoxin: CO_2 oxidoreductase from *Clostridium pasterianum*, a molybdenum iron-sulfur-protein, *Eur. J. Biochem.*, 85, 125, 1978.
181. **Enoch, H. G. and Lester, R. L.,** The purification and properties of formate dehydrogenase and nitrate reductase from *Escherichia coli, J. Biol. Chem.*, 250, 6693, 1975.
182. **Muller, W., Willnow, P., Ruschig, U., and Hopner, T.,** Formate dehydrogenase from *Pseudomonas oxalaticus; Eur. J. Biochem.*, 83, 485, 1978.
183. **Egorov, A. M., Avilova, T. V., Dikov, M. M., Popov, V. O., Rodionov, Y. V., and Berezin, I. V.,** NAD-dependent formate dehydrogenase from methylotrophic bacterium, strain 1, *Eur. J. Biochem.*, 99, 569, 1979.
184. **Van Dijken, J. P., Oostra-Demkes, G. J., Otto, R., and Harder, W.,** S-Formylglutathione: the substrate for formate dehydrogenase in methanol-utilizing yeasts, *Arch. Microbiol.*, 111, 77, 1976.
185. **Schutte, H., Flossdorf, J., Sahm, H., and Kula, M. R.,** Purification and properties of formaldehyde dehydrogenase and formate dehydrogenase from *Candida boidinii, Eur. J. Biochem.*, 62, 151, 1976.
186. **Egorov, A. M., Tishkov, V. I., Avilova, T. V., and Popov, V. O.,** S-Formylglutathione as a substrate of bacterial formate dehydrogenase, *Biochem. Biophys. Res. Commun.*, 104, 1, 1982.
187. **Leadbetter, E. R. and Foster, J. W.,** Bacterial oxidation of gaseous alkanes, *Arch. Mikrobiol.*, 35, 92, 1960.
188. **Hou, C. T., Patel, R. N., Laskin, A. I., Marczak, I., and Barnabe, N.,** Microbial oxidation of gaseous hydrocarbons: production of alcohols and methyl ketones from their corresponding *n*-alkanes by methylotrophic bacteria, *Can. J. Microbiol.*, 27, 107, 1981.
189. **Patel, R. N., Hou, C. T., Laskin, A. I., Derelanko, P., and Felix, A.,** Oxidation of secondary alcohols to methyl ketones by yeasts, *Appl. Environ. Microbiol.*, 38, 219, 1979.
190. **Hou, C. T., Patel, R. N., Laskin, A. I., Barnabe, N., and Marczak, I.,** Substrate specificity and stereospecificity of nicotinamide adenine dinucleotide-linked alcohol dehydrogenases from methanol-grown yeasts, *Appl. Environ. Microbiol.*, 41, 829, 1981.

191. **Hou, C. T., Patel, R. N., Laskin, A. I., Barnabe, N., and Marczak, I.,** Identification and purification of a nicotinamide adenine dinucleotide-dependent secondary alcohol dehydrogenase from C_1-utilizing microbs, *FEBS Letters*, 101, 179, 1979.
192. **Hou, C. T., Patel, R. N., Barnabe, N., and Marczak, I.,** Stereospecificity and other properties of a novel secondary-alcohol dehydrogenase, *Eur. J. Biochem.*, 119, 359, 1981.
193. **Patel, R. N., Hou, C. T., Laskin, A. I., Derelanko, P., and Felix, A.,** Microbial production of methyl ketones: purification and properties of a secondary alcohol dehydrogenase from yeast, *Eur. J. Biochem.*, 101, 401, 1979.
194. **Patel, R. N., Hou, C. T., Laskin, A. I., and Derelanko, P.,** Microbial production of methylketones: properties of purified yeast secondary alcohol dehydrogenase, *J. Appl. Biochem.*, 3, 218, 1981.
195. **Dickinson, F. M., and Dalziel, K.,** Substrate specificity and stereospecificity of alcohol dehydrogenase, *Nature (London)*, 214, 31, 1967.
196. **Peterson, J. A., Basu, D., and Coon, M. J.,** Enzymatic w-oxidation. I. Electron carriers in fatty acid and hydrocarbon hydroxylation, *J. Biol. Chem.*, 241, 5162, 1966.
197. **Anthony, C.,** The prediction of growth yields in methylotrophs, *J. Gen. Microbiol.*, 104, 91, 1978.
198. **Wadzinski, A. M. and Ribbons, D. M.,** Utilization of acetate by *Methanomonas methanooxidans*, *J. Bacteriol.*, 123, 380, 1975.
199. **Eccleston, M. and Kelly, D. P.,** Assimilation and toxicity of some exogenous C_1 compounds, alcohols, sugars and acetate in the methane oxidising bacterium, *Methylococcus capsulatus*, *J. Gen. Microbiol.*, 75, 211, 1973.
200. **Foster, J. W.,** Bacterial oxidation of hydrocarbons in *Oxygenases*, Hayaishi, O., Ed., Academic Press, New York, 1962, 241.
201. **Perry, J. J.,** Microbial cooxidations involving hydrocarbons, *Microbiol. Review*, 43, 59, 1979.
202. **Quayle, J. R.,** Historical Prospectives, Proc. of the Institute of Petroleum Microbiology Group Meeting on Hydrocarbons in Biotechnology p 1, University of Kent, Canterbury, September, 1979.
203. **Higgins, I. J., Hammond, R. C., Sariaslani, F. S., Best, D., Davies, M. M., Tryhorn, S. E., and Taylor, F.,** Biotransformation of hydrocarbons and related compounds by whole organism suspension of methane-grown *Methylosinus trichosporium* OB3b, *Biochem. Biophys. Res. Commun.*, 89, 671, 1979.
204. **Hou, C. T.,** Microbial transformation of important industrial hydrocarbons in *Microbial Transformation of Physiologically Active Compounds*, Rossaza, J., Ed., CRC Press, 1982, 81.
205. **Stirling, D. I. and Dalton, H.,** The fortuitous oxidation and cometabolism of various carbon compounds by whole-cell suspensions of *Methylococcus capsulatus* Bath, *FEMS Microbiol. Lett.*, 5, 315, 1979.
206. **Stirling, D. I. and Dalton, H.,** Fortuitous oxidation by methane-utilizing bacteria, *Nature (London)*, 291, 169, 1981.
207. **Higgins, I. J., Best, D. J., and Hammond, R. C.,** New findings in methane-utilizing bacteria, highlight their importance in the biosphere and their commercial potential, *Nature (London)*, 286, 561, 1980.
208. **Higgins, I. J., Best, D. J., and Hammond, R. C.,** Reply to Stirling and Dalton, *Nature (London)*, 291, 169, 1981.
209. **Higgins, I. J.,** Respiration in methylotrophic bacteria, in *Diversity of Bacterial Respiratory Systems*, Vol. 1, Knowles, C. J., Ed., CRC Press, Boca Raton, FL, 1980, 187.
210. **Davey, J. F. and Milton, J. R.,** Cytochromes of two methane-utilizing bacteria, *FEBS Lett.*, 37, 335, 1973.
211. **Weaver, T. L. and Dugan, P. R.,** Serine transhydroxymethylase isoenzymes from a facultative methylotroph, *J. Bacteriol.*, 124, 985, 1975.
212. **Monosov, E. Z. and Netrusov, A. I.,** Localization of energy generators in methane oxidizing bacteria, *Microbiology (U.S.S.R.)*, 45, 598 1976.
213. **Higgins, I. J., Knowles, C. J., and Gonge, G. M.,** *Microbial Production and Utilization of Gases (H_2, CH_4, CO)*, Schlegel, H. G., Gottschalk, G., and Pfenning, N., Eds., Erich Goltze, K. G. Gottingen, 1976, 389.
214. **Tonge, G. M., Knowles, C. J., Harrison, D. E. F., and Higgins, I. J.,** Metabolism of one-carbon compounds: cytochromes of methane- and methanol-utilizing bacteria, *FEBS Lett.*, 44, 106, 1974.
215. **Ferenci, T.,** Oxygen metabolism in *Pseudomonas methanica*, *Arch. Microbiol.*, 108, 217, 1976.
216. **Anthony, C.,** Microbial metabolism of C_1-compounds: the cytochromes of *Pseudomonas* AM1, *Biochem. J.*, 146, 289, 1975.
217. **Widdowson, D. and Anthony, C.,** The microbial metabolism of C_1 compounds: the electron-transport chain of *Pseudomonas* AM1, *Biochem. J.*, 152, 349, 1975.
218. **Drabikowska, A. K.,** The respiratory chain of a newly isolated *Methylomonas* Pl1, *Biochem J.*, 168, 171, 1977.
219. **Patel, R. N., Hou, C. T., and Felix, A.,** Purification and properties of cytochrome C from *Methylomonas methanica*, *J. Gen. Appl. Microbiol.*, 25, 197, 1979.
220. **Cross, A. R. and Anthony, C.,** The purification and properties of the soluble cytochromes C of the obligate methylotroph *Methylophilus methylotrophus*, *Biochem. J.*, 192, 421, 1980.

221. **O'Keeffe, D. T. and Anthony C.**, The two cytochromes C in the facultative methylotroph *Pseudomonas* AM1, *Biochem. J.*, 192, 411, 1980.
222. **Anthony, C.**, *Microbial Growth on C_1 Compounds*, Dalton, H., Ed., Heyden, London, 1981, 220.
223. **Ohta, S. and Tobari, J.**, Two cytochromes C of *Methylomonas* J., *J. Biochem.*, 90, 215, 1981.
224. **Beardmore-Gray, M., O'Keeffe, D. T., and Anthony, C.**, The autoreducible cytochromes C of the methylotrophs *Methylophilus methylotrophus* and *Pseudomonas* AM1, *Biochem. J.*, 207, 161, 1982.
225. **Mitchell, P.**, Chemiosmotic coupling in oxidative and photosynthetic phosphorylation, *Biol. Rev.*, 41, 445, 1966.
226. **Tonge, G. M., Drozd, J. W., and Higgins, I. J.**, Energy coupling in *Methylosinus trichosporium*, *J. Gen. Microbiol.*, 99, 229, 1977.
227. **Netrusov, A. I., Rodionov, Y. V., and Kondratieva, E. N.**, ATP-generation coupled with C_1-compound oxidation by methylotrophic bacterium *Pseudomonas* sp. 2, *FEBS Lett.*, 76, 56, 1977.
228. **O'Keeffe, D. and Anthony, C.**, The microbial metabolism of C_1-compounds. The stoichiometry of respiration-driven proton translation in *Pseudomonas* AM1 and in a mutant lacking cytochrome C, *Biochem. J.*, 170, 561, 1978.
229. **Dawson, M. J. and Jones, C. W.**, Respiration-linked proton translocation in the obligate methylotroph *Methylophilus methylotrophus Biochem. J.*, 194, 915, 1981.
230. **Matsumoto, T.**, Methylamine dehydrogenase of *Pseudomonas* sp. J. Purification and properties, *Biochem. Biophys. Acta*, 522, 291, 1978.
231. **Metha, R. J.**, Methylamine dehydrogenase from the obligate methylotroph *Methylomonas methylovora*, *Can. J. Microbiol.*, 23, 402, 1977.
232. **Eady, R. R. and Large, P. J.**, Purification and properties of an amine dehydrogenase from *Pseudomonas* AM1 and its role in growth on methylamine, *Biochem. J.*, 106, 245, 1968.
233. **Matsumoto, T., Hiraoka, B. Y., and Tobari, J.**, Methylamine dehydrogenase of *Pseudomonas sp.* J. Isolation and properties of the subunits, *Biochim. Biophys. Acta*, 522, 303, 1978.
234. **Matsumoto, T. and Tobari, J.**, Methylamine dehydrogenase of *Pseudomonas* sp. J. Reconstitution from its subunits, *J. Biochem.*, 83, 1591, 1978.
235. **Matsumoto, T. and Tobari, J.**, Methylamine dehydrogenase of *Pseudomonas* sp. J. and *Pseudomonas* AM1. Study on subunit structure by dimethyl suberimidate, *J. Biochem.*, 84, 461, 1978.
236. **Pollock, R. J. and Hersh, L. B.**, *N*-methylglutamate synthetase. Purification and properties of the enzyme, *J. Biol. Chem.*, 246, 4737, 1971.
237. **Hersh, L. B., Peterson, J. A., and Thompson, A. A.**, An *N*-methylglutamate dehydrogenase from *Pseudomonas* MA, *Arch. Biochem. Biophys.*, 150, 219, 1972.
238. **Bamforth, C. W. and Large, P. J.**, Solubilization, partial purification and properties of *N*-methylglutamate dehydrogenase from *Pseudomonas aminovorans*, *Biochem. J.*, 161, 357, 1977.
239. **Bamforth, C. W. and Large, P. J.**, The molecular size of *N*-methylglutamate dehydrogenase of *Pseudomonas aminovorans*, *Biochem. J.*, 167, 509, 1977.
240. **Netrusov, A. I.**, NAD-dependent *N*-methylglutamate dehydrogenase, a new enzyme of methylamine metabolism in microorganisms, *Microbiology*, 44, 552, 1975.
241. **Loginova, N. V., Shishkina, V. N., and Trotsenko, Yu. A.**, Primary metabolic pathways of methylated amines in *Hyphomicrobium vulgare*, (transl) *Mikrobiologiya*, 45, 34, 1976.
242. **Lim, M. C. M. and Wagner, C.**, Purification and characterization of *N*-methylamine dehydrogenase, *J. Biol. Chem.*, 250, 3746, 1975.
243. **Kung. H.-F. and Wagner, C.**, γ-Glutamyl methylamine. A new intermediate in the metabolism of methylamine, *J. Biol. Chem.*, 244, 4136, 1969.
244. **Levitch, M. E.**, In Microbial growth on C_1-compounds, Skryabin, G. K., Ivanov, M. V., Kondratjeva, E. N., Zavarzin, G. A., Trotsenko, Yu. A., and Nesterov, A. I., Eds., pp 41, 1977, U.S.S.R., Acad. Sci. Puschino.
245. **Loginova, N. V.**, in *Microbial Growth on C_1-Compounds*, Skryabin, G. K., Ivanov, M. V., Kondratjeva, E. N., Zavarzin, G. A., Trotsenko, Yu. A., and Nesterov, A. I., Eds., U.S.S.R. Academy of Sciences, Puschino, 1977, 39.
246. **Loginova, N. V. and Trotsenko, Yu. A.**, Pathways of oxidation and assimilation of methylated amines in *Arthrobacter globiformis*, *Microbiology*, 44, 892, 1976.
247. **Meiberg, J. B. M. and Harder, W.**, Aerobic and anaerobic metabolism of trimethylamine, dimethylamine and methylamine in *Hyphomicrobium* X, *J. Gen. Microbiol.*, 106, 265, 1978.
248. **Bamforth, C. W. and O'Connor, M. L.**, The isolation of pleiotropic mutants of *Pseudomonas aminovorans* deficient in the ability to grow on methylamine and an examination of their enzymic constitution, *J. Gen. Microbiol.*, 110, 143, 1979.
249. **Colby, J. and Zatman, L. J.**, The purification and properties of a bacterial trimethylamine dehydrogenase, *Biochem. J.*, 121, 9, 1971.

250. **Large, P. J., Boulton, C. A., and Crabbe, M. J. C.,** The reduced nicotinamide-adenine dinucleotide phosphate- and oxygen-dependent *N*-oxygenation of trimethylamine by *Pseudomonas aminovorans*, *Biochem. J.*, 128, 137, 1972.
251. **Myers, P. A. and Zatman, L. J.,** The metabolism of trimethylamine *N*-oxide by *Bacillus* PM6, *Biochem. J.*, 121, 10, 1971.
252. **Large, P. J.,** Non-oxidative demethylation of trimethylamine *N*-oxide by *Pseudomonas aminovorans*, *FEBS Lett.*, 18, 297, 1971.
253. **Colby, J. and Zatman, L. J.,** Purification and properties of the trimethylamine dehydrogenase of *Bacterium* 4B6, *Biochem. J.* 143, 555, 1974.
254. **Steenkamp, D. J. and Mallinson, J.,** Trimethylamine dehydrogenase from a methylotrophic bacterium I. Isolation and steady-state kinetics, *Biochim. Biophys. Acta*, 429, 705, 1976.
255. **Steenkamp, D. J. and Gallup, M.,** The natural flavoprotein electron acceptor of trimethylamine dehydrogenase, *J. Biol. Chem.*, 253, 4086, 1978.
256. **Large, P. J. and McDougall, M.,** An enzymic method for the microestimation of trimethylamine, *Anal. Biochem.*, 64, 304, 1975.
257. **Steenkamp, D. J., Kenney, W. C., and Singer, T. P.,** A novel type of covalently bound co-enzyme in trimethylamine dehydrogenase, *J. Biol. Chem.*, 253, 2812, 1978.
258. **Steenkamp, D. J., McIntire, W., and Kenney, W. C.,** Structure of the covalently bound coenzyme of trimethylamine dehydrogenase. Evidence for a 6-substituted flavin, *J. Biol. Chem.*, 253, 2818, 1978.
259. **Steenkamp, D. J. and Singer, T. P.,** On the presence of a novel covalently bound oxidation-reduction cofactor, iron and labile sulfur in trimethylamine dehydrogenase, *Biochem. Biophys. Res. Commun.*, 71, 1289, 1976.
260. **Steenkamp, D. J., Singer, T. P., and Beinert, H.,** Purification of the iron-sulfur cluster and of the covalently bound coenzyme of trimethylamine dehydrogenase in catalysis, *Biochem. J.*, 169, 361, 1978.
261. **Hill, C. L., Steenkamp, D. J., Holm, R. H., and Singer, T. P.,** Identification of the iron-sulfur cluster in trimethylamine dehydrogenase, *Proc. Natl. Acad. Sci. U.S.A.*, 74, 547, 1977.
262. **Steenkamp, D. J. and Beinert, H.,** Mechanistic studies on the dehydrogenases of methylotrophic bacteria. I. The influence of substrate binding to reduced trimethylamine dehydrogenase on the intramolecular electron transfer between its prosthetic groups, *Biochem. J.*, 207, 233, 1982.
263. **Steenkamp, D. J. and Beinert, H.,** Mechanistic studies on the dehydrogenases of methylotrophic bacteria. II. Kinetic studies on the intramolecular electron transfer in trimethylamine and dimethylamine dehydrogenase, *Biochem. J.*, 207, 241, 1982.
264. **Eady, R. R. and Large, P. J.,** Bacterial oxidation of dimethylamine, a new monooxygenase reaction, *Biochem. J.*, 111, 37, 1969.
265. **Brook, D. F. and Large, P. J.,** Inhibition by carbon monoxide of the secondary-amine mono-oxygenase of *Pseudomonas aminovorans* and the photochemical action spectrum for its reversal, *Eur. J. Biochem.*, 55, 601, 1975.
266. **Brook, D. F. and Large, P. J.,** A steady-state kinetic study of the reaction catalyzed by the secondary-amine mono-oxygenase of *Pseudomonas aminovorans*, *Biochem. J.*, 157, 197, 1976.
267. **Eady, R. R., Jarman, T. R., and Large, P. J.,** Microbial oxidation of amines. Partial purification of a mixed-function secondary-amine oxidase system from *Pseudomonas aminovorans* that contains an enzymically active cytochrome-P-420-type haemoprotein, *Biochem. J.*, 125, 449, 1971.
268. **Meiberg, J. B. M. and Harder, W.,** Dimethylamine dehydrogenase from *Hyphomicrobium* X: purification and some properties of a new enzyme that oxidizes secondary amines, *J. Gen. Microbiol.*, 115, 49, 1979.
269. **Steenkamp, D. J.,** Identification of the prosthetic groups of dimethylamine dehydrogenase from *Hyphomicrobium* X, *Biochem. Biophys. Res. Commun.*, 88, 244, 1979.
270. **Panganiban, A. t., Patt, T. E., Hart, W., and Hanson, R. S.,** Oxidation of methane in the absence of oxygen in lake water samples, *Appl. Environ. Microbiol.*, 37, 303, 1979.
271. **Daniels, L., Fuchs, G., Thauer, R. K., and Zeikus, J. G.,** Carbon monoxide oxidation by methanogenic bacteria, *J. Bacteriol.*, 132, 118, 1977.
272. **Weimer, P. J. and Zeikus, J. G.,** One carbon metabolism in methanogenic bacteria: cellular characterization and growth of *Methanosarcina barkeri*, *Arch. Microbiol.*, 119, 49, 1978.
273. **Zeikus, J. G., Lynd, L. H., Thompson, T. E., Krzycki, J. A., Weimer, P. J., and Hegge, P. W.,** Isolation and characterization of a new, methylotrophic, acidogenic anaerobe, the Marburg strain, *Current Microbiol.*, 3, 381, 1980.
274. **Brown, C. M., Macdonald-Brown, D. S., and Meers, J. L.,** Physiological aspects of microbial inorganic nitrogen metabolism, *Adv. Microbial. Physiol.*, 11, 1, 1974.
275. **Drozd, J. W., Linton, J. D., Downs, J., Stephenson, R., Bailey, M. L., and Wren, S. J.,** Growth energetics in methylotrophic bacteria, in *Microbial Growth on C_1-Compounds*, Skryabin, G. K., Ivanov, M. V., Kondratjeva, E. N., Zavarzin, G. A., Trotsenko, Yu. A., and Nesterov, A. I., Eds., U.S.S.R. Academy of Sciences, Pushchino, 1977, 91.

276. **O'Neill, J. G. and Wilkinson, J. F.,** Oxidation of ammonia by methane-oxidising bacteria and the effects of ammonia on methane oxidation, *J. Gen. Microbiol.*, 100, 407, 1977.

277. **Dalton, H.,** Ammonia oxidation by the methane oxidising bacterium *Methylococcus capsulatus* strain Bath, *Arch. Microbiol.*, 114, 273, 1977.

278. **Sokolov, I. G., Romanovskaya, V. A., Shkurko, Y. B., and Malashenko, Y. R.,** Comparative characterisation of the enzyme systems of methane-utilizing bacteria that oxidize NH_2OH and CH_3OH, *Microbiology (U.S.S.R.)*, 49, 142, 1980.

279. **Coty, V. F.,** Atmospheric nitrogen fixation by hydrocarbon-oxidizing bacteria, *Biotech. Bioeng.*, 9, 25, 1967.

280. **de Bont, J. A. M.,** Hydrogenase activity in nitrogen fixing, methane-oxidising bacteria, *Antonie van Leeuwenhoek J. Microbiol. Serol.*, 42, 255, 1976.

281. **De Bont, J. A. M.,** Nitrogen fixation by methane-utilizing bacteria, *Antonie van Leeuwenhoek, J. Microbiol. Serol.*, 42, 245, 1976.

282. **de Bont, J. A. M. and Mulder, E. G.,** Invalidity of the acetylene reduction assay in alkane-utilizing, nitrogen-fixing bacteria, *Appl. Environ. Microbiol.*, 31, 640, 1976.

283. **Dalton, H. and Whittenbury, R.,** The acetylene reduction technique as an assay for the nitrogenase activity in the methane oxidizing bacterium *Methylococcus capsulatus* strain Bath, *Arch. Microbiol.*, 109, 147, 1976.

284. **Taylor, I. J. and Anthony, C.,** A biochemical basis for obligate methylotrophy: properties of a mutant of *Pseudomonas* AM1 lacking 2-oxoglutarate dehydrogenase, *J. Gen. Microbiol.*, 93, 259, 1976.

285. **Shishkina, V. N. and Trotsenko, Y. A.,** Multiple enzymic lesions in obligate methanotrophic bacteria, *FEMS Microbiol. Lett.*, 13, 237, 1982.

286. **Green, P. N. and Bousfield, I. J.,** Emendation of *Methylobacterium* Patt, Cole, and Hanson 1976; *Methylobacterium rhodinum* (Huemann 1962) comb. nov. corrig.; *Methylobacterium radiotolerans* (Ito and Iizuka 1971) comb nov. corrig., and *Methylobacterium mesophilicum* (Austin and Goodfellow 1979) comb. nov, *Int. J. System. Bacteriol.*, 33, 875, 1983.

287. **Stanley, S. H., Prior, S. D., Leak, D. J., and Dalton, H.,** Copper stress underlies the fundamental change in intracellular location of methane mono-oxygenase in methane-oxidizing organisms: studies in batch and continuous cultures, *Biotechnol. Lett.*, 5, 487, 1983.

288. **Hou, C. T., Patel, R. N., Laskin, A. I., Barnabe, N., and Barist, I.,** Purification and properties of a NAD-linked 1,2-propanediol dehydrogenase from propane-grown *Pseudomonas fluorescens*, NRRL B-1244, *Arch. Biochem. Biophys.*, 223, 297, 1983.

Chapter 2

MICROBIOLOGY AND BIOCHEMISTRY OF METHYLOTROPHIC YEASTS

Yoshiki Tani

TABLE OF CONTENTS

I. INTRODUCTION

Until the first isolation of *Kloeckera* sp. No. 2201 as a methanol-utilizing yeast in 1969 by Ogata et al.,[1] no eukaryote was recognized to be able to grow on reduced C_1-compounds, e.g., methane and methanol, as the sole source of carbon and energy. In 1948, Wickerham and Burton[2] classified methanol to a group of compounds, which were not utilized by yeast, through the test for the carbon assimilation with 100 strains of yeast representing 22 genera. Since then, methanol had been neglected as a possible growth substrate for yeasts. Recovery of yeasts from the culture broth is significantly easier and cheaper than that of bacteria due to the larger diameter of the yeast cell. As single-cell protein, yeast is psychologically more acceptable for human consumption because of the experience of using the cells as a food supplement. For these reasons, investigations on methanol-utilizing yeasts have attracted much attention in applied and fundamental fields of microbiology during these $1^1/_2$ decades.[3-6]

Methanol-utilizing yeasts reported so far are restricted to seven genera and are all facultative methylotrophs. With respect to their mode of methanol metabolism, these yeasts form a homogeneous group in contrast to methylotrophic bacteria.[7] They oxidize methanol to formaldehyde with alcohol oxidase and formaldehyde to CO_2 through formate with NAD-dependent dehydrogenases, and they might assimilate methanol through the same formaldehyde-fixation system.

In this chapter, microbiological and metabolic features of the methanol-utilizing yeasts will be described. Other methylotrophic eukaryotes such as methylated amine- and methane-utilizing yeasts and methanol-utilizing molds are also included.

II. MICROBIOLOGY OF METHANOL-UTILIZING YEASTS

A. Isolation

Ogata et al.[1] first reported the isolation of methanol-utilizing yeasts from natural sources.* Among them, strain No. 2201, which showed good growth on methanol medium, was selected and characterized.[9] The strain was identified to be a new strain related to the genus *Kloeckera* and thereafter was reidentified as *Candida boidinii*.[10] A variety of carbon compounds are utilized by the yeast, but formate, formaldehyde and methylamine are not. The cell yield is 2.25 and 4.6 g/ℓ of media containing 1 and 3% methanol, respectively. The cell compositions of the yeast, which are crude protein, 45.30%; crude fat, 5.24%; crude

* Before the first description of methylotrophic yeast, Aroj et al. showed the possibility of acclimatization of *Candida* yeast culture towards methanol.[8] The yeast consumed methanol in the later stages of generations when the active culture became resistant to methanol doses. This was not, however, followed by other researchers.

fiber, 4.84%; nitrogen-free extract, 33.67%; ash, 6.90%; and nucleic acid, 5.4%, rated methanol-utilizing yeast as a possible source of single-cell protein.[11]

Since the first isolation, a number of yeasts have been found to be able to grow on methanol as the sole source of carbon and energy. Hazeu et al.[12] tested preserved strains including all type strains of yeast species in the collection of the Centralbureau voor Schimmelcultures, Delft, for methanol-utilizing ability. All *Candida boidinii, Pichia pinus*, and *Hansenula capsulata* strains tested consistently grow on methanol. They also pointed out that the majority of the methanol-utilizing yeast strains are isolated from the bark of trees or from insects living on trees, and that the number of these yeasts might be enhanced in this ecological niche by the presence of compounds such as lignin, being rich to methoxy group. Ogata et al.[13] showed another example of distribution of methanol-utilizing yeasts in nature. About 55 cultures of yeast were isolated from 600 samples of a wide variety of natural sources by enrichment cultures on methanol medium. Among the sources, wood and flower are good samples for isolation of such yeasts.

The optimal conditions for isolation of methanol-utilizing yeasts are also discussed.[14] The reasons for successful isolation might be attributed to the following: the use of lower concentration of methanol, 0.1 to 0.5% (w/v), to prevent the growth inhibition; the addition of vitamin mixtures, especially biotin and thiamin; the addition of penicillin G and D-cycloserine; the adjustment of pH of medium to 4.5, and the use of soil samples rich in organic matter as inoculum.

Many yeasts have been isolated for further selection of strains with higher growth rate. Growth temperature is also an essential property of microorganisms in the cost of single-cell protein production. Levine and Cooney[15] isolated a thermotolerant methanol-utilizing yeast, *Hansenula polymorpha* DL-1, by a continuous enrichment culture at 37°C. The yeast can grow in a chemostat at temperature up to 50°C, with strong growth at 45°C. A screening method to select the thermophilic yeasts of high growth rate was developed;[16] soil samples collected from tropical regions are rich in thermophilic yeasts. An ordinary enrichment culture using a test tube is more effective than a continuous enrichment culture using a minifermentor because many prominent yeasts are lost through the latter culture. The screening test by comparing the growth rate of the culture in a flask also results in loss of the type of yeasts which grows at low growth rate in batch culture and at high growth rate in continuous culture. A new fermentor with a high oxygen transfer capacity was designed and employed for the final screening.[17] A thermophilic yeast thus selected, *Pichia methanothermo*, yields 4.12 g cells per liter per hour with continuous culture in a medium containing 73 g of methanol per liter and with a dilution rate of 0.146/hr, temperature of 40°C, and cell density of 28.2 g/ℓ.[18]

Hansenula polymorpha strain isolated by Pal and Hamdan[19] is another type of methanol-utilizing yeast having favorable properties in single-cell protein production. The yeast was isolated from a mixed culture obtained from seawater and is halophilic thermotolerant. It grows on methanol at 25 to 42°C, pH 3.5 to 6.7, and in a medium compounded with 75% seawater.

B. Taxonomy

About 50 strains of yeasts are reported as methanol-utilizer. These species are involved in seven genera which include both ascomycetous and asporogenous yeasts. All can utilize carbon compounds other than methanol and are facultative methylotrophs. Multipolar budding and a requirement for biotin and/or thiamin are common features of many of these yeasts. These peculiar characteristics of methanol-utilizing yeasts may lead to a classification system forming a new group of yeasts.

A taxonomic study on methanol-utilizing yeasts were demonstrated by Lee and Komagata.[10] The DNA base composition (GC content of DNA), ubiquinone system, proton magnetic resonance spectra of cell-mannans and electrophoretic patterns of alcohol oxidase of these yeasts were examined in addition to the morphological characterization. The GC content of DNA is distributed in a broad range from 29.2 to 50.2%. The alcohol oxidase is divided into three types by the zymograms. The ubiquinone system and proton magnetic resonance spectra are characteristic for each species. On the basis of the chemotaxonomic characteristics, methanol-utilizing yeasts are divided into four groups:

1. Group I. This group consists only of *Candida boidinii*, which has a lower GC content of DNA than any other yeasts.
2. Group II. Almost all of the yeasts belong to this group. The GC content of DNA of this group has a broad distribution from 34.6 to 50.2%. The ubiquinone system is Q7 and the proton magnetic resonance spectrum shows main peaks with 5.05 and 5.19.
3. Group III. Only *Pichia cellobiosa* belongs to this group, which has a characteristic ubiquinone system of Q8 and Q7.
4. Group IV. *Hansenula capsulata* and *Pichia pastoris* belongs to this group, which has a ubiquinone system of Q8. The proton magnetic resonance spectrum and the electrophoretic pattern of alcohol oxidase are different from those of the other groups.

Methanol-utilizing yeasts that were characterized are listed in Table 1.

C. Growth Characteristics

Growth inhibition of methanol at high concentration is a general problem in the cultivation of methylotrophs. The growth of methanol-utilizing yeasts is usually affected at the methanol concentration of 0.5% (w/v). The dependence of the specific growth rate of *Candida boidinii* on methanol concentration was observed to follow Monod-type kinetics in the range of 0 to 0.64%.[32] In the case of *Torulopsis methanolovescens* the growth is not affected at the concentration of 1%.[33] On increasing the initial methanol concentration, the maximum dry cell weight increases, the overall growth yield decreases, and the lag time is prolonged.

A specific technique of methanol feeding in cultivation of *Candida boidinii* was developed.[34] In the technique, extended culture, methanol is added to a batch process under controlled conditions. The control loop is based on the measurement of methanol in the exhaust air by means of flame-ionization detector. The results from the extended batch fermentation at different methanol concentrations showed that an increase in growth activity up to 0.2% methanol is followed by a decline due to the substrate inhibition. For methanol feeding with less growth inhibition, an apparatus for automatic drawing of the growth curve of yeast was also constructed.[35] In the apparatus the integration of the pH-controller operating time or the consumption of the pH-controlling agent is proportional to the cell concentration in culture broth.

Formaldehyde and formate, the oxidative products of methanol, are more toxic to growth than methanol, approximately 100 and 10 times, respectively.[32] Therefore, the cell must be very well adapted to methanol to overcome transient accumulation of formaldehyde and formate in the medium. Continuous culture is shown to be advantageous for this purpose because well defined culture conditions can be held for a long enough time for sufficient adaptation.[36]

The relatively high vapor pressure of methanol causes an additional problem: methanol is stripped off with aeration when in high concentration in the culture broth. *Pichia methanothermo* was cultivated by supplying methanol in vapor form or in liquid form through a programmed automatic pump.[37] The lag phase of culture was shortened by about 10 hr in flask culture and 2 to 4 hr in fermentor culture.

Involvement of an oxygen-dependent alcohol oxidase in the primary step in methanol oxidation of methanol-utilizing yeasts might require higher oxygen concentration in the

Table 1
METHANOL-UTILIZING YEASTS

Strain	Origin[a]	Characteristic	Group[b]	Ref.
Ascomycetous yeasts				
Schizosaccharomyces spp. 7Av, 12Av	A			20
Saccharomyces sp. H-1	A	Biotin-requiring		21
Pichia methanolica	A	Maximum growth temperature, 38—40°C		22
Pichia pastoris	A		IV	22
Pichia pastoris	B	CBS 704	IV	12, 14
Pichia pinus	B	CBS 744	II	12, 14
Pichia trehalophila	B	CBS 5361	II	12, 14
Pichia lindnerii	A	Biotin-requiring	II	23
Pichia methanothermo	A	Optimum growth temperature, 37—40°C		18
Pichia cellobiosa	A	Biotin-requiring	III	24
Hansenula wickerhamii	B	CBS 4307	II	14
Hansenula capsulata	B	CBS 1993	IV	14
Hansenula glucozyma	B	CBS 5766	II	12
Hansenula henricii	B	CBS 5765	II	12, 14
Hansenula minuta	B	CBS 1708	II	12
Hansenula mollischiana	B			12
Hansenula nonfermentans	B	CBS 5764	II	12, 14
Hansenula philodendra	B	CBS 6075	II	12
Hansenula polymorpha	B	CBS 4732	II	12, 14
Hansenula polymorpha DL-1	A	Thermotolerant, biotin-requiring, growth stimulation by thiamin	II	15
Hansenula ofunaensis	A			25
Hansenula polymorpha	A	Holophilic, thermotolerant, biotin- and thiamin-requiring		19
Asporogenous yeasts				
Kloeckera sp. (*Candida boidinii*)[c] No. 2201	A	Growth stimulation by biotin and thiamin	I	1
Candida spp. N-16, N-17	A	Biotin-requiring		21
Candida methanolica(boidinii)[c]	A	Growth stimulation by biotin	I	26
Candida boidinii	B	CBS 2428	I	12, 14
Canida boidinii	A	Biotin-requiring	I	27
Candida boidinii	A		I	22
Candida boidinii	A	Thermotolerant	I	28
Candida methanophilum	A	Growth stimulation by biotin		29
Candida cariosilignicola	A	Biotin-requiring	II	24
Candida succiphila	A	Biotin-requiring	II	24
Torulopsis methanolovescens	A	Biotin- and thiamin-requiring	II	26
Torulopsis glabrata	A		II	30

Table 1 (continued)
METHANOL-UTILIZING YEASTS

Strain	Origin[a]	Characteristic	Group[b]	Ref.
Torulopsis nemodendra	B	CBS 6280	II	12
Torulopsis nitratophila	B	CBS 2027	II	12
Torulopsis pinus	B	CBS 970	II	12
Torulopsis molischana	B		II	12
Torulopsis methanosorbosa	A	Growth stimulation by biotin and thiamin		31
Torulopsis methanodomercqii	A	Growth stimulation by biotin and thiamin		31
Torulopsis nagoyaensis	A			25

[a] Origin: A, isolated by enrichment techniques; B, type strain in culture collection.
[b] Group: referred to Lee and Komagata.[10]
[c] Reidentified by Lee and Komagata.[10]

culture with methanol than in cultures with other carbon sources. The respiration with methanol as substrate gave substrate constants for oxygen between 0.5 and 1.3 mg O_2 per liter, whereas oxygen uptake due to a type of endogenous respiration or that with ethanol was down to about 0.15 mg/ℓ.[38] The dissimilation of methanol by alcohol oxidase is also reflected in the growth yield of the yeasts. The maximal yield of *Hansenula polymorpha,* for example, was reported to be low (0.38 cells per gram methanol).[39]

The values of the growth yield coefficient (Y_s) of 0.3 and the respiration quotient (RQ) of 0.5 of growth on methanol of *Candida boidinii* are lower than those on glucose. Based on the values, it is calculated that 38% of methanol consumed is incorporated into cell materials and 62% is oxidized to CO_2.[40]

In comparison to bacteria, the low growth yields and low growth rates during growth of yeasts with methanol are major disadvantages for industrial applications. Enhanced overall growth yields and the ability to utilize mixed substrates at dilution rates higher than the corresponding maximal growth rates attainable for each singular substrate are known for some facultative autotrophic bacteria. Egli et al.[41,42] investigated extensively the growth of *Hansenula polymorpha* and *Kloeckera* sp. No. 2201 with a mixture of methanol and glucose. The yeasts can assimilate methanol and glucose simultaneously in carbon-limited chemostat culture. At low dilution rate, the two carbon sources are assimilated with approximately the same efficiency as during single substrate limited growth and therefore no improved overall growth yields are obtained. During the increase of the dilution rate up to 0.30/hr for *Hansenula polymorpha* and 0.26/hr for *Kloeckera* sp. No. 2201, they assimilate methanol and glucose simultaneously with significantly higher growth rate than the maximum specific growth rate obtained with methanol as the sole carbon source.

D. Derivation of Mutants

A mutant of methanol-utilizing yeasts was first derived with *Candida boidinii* by ordinary process using *N*-methyl-*N'*-nitro-*N*-nitrosoguanidine as mutagen.[43] The mutant is unable to grow on methanol as the sole source of carbon and energy, lacking the activity of alcohol oxidase. A procedure was developed to isolate mutants altered in dissimilatory enzymes.[44] The method involves permeation of cells by the treatment with chitosan and staining of colonies by the specific enzyme assay, coupled with visible color reactions for the qualitative direct identification of alcohol oxidase, alcohol dehydrogenase, and formaldehyde dehydrogenase. In the combination with mutation with *N*-methyl-*N'*-nitro-*N*-nitrosoguanidine, this method was applied successfully to isolate mutants of *Hansenula polymorpha*.

Mutants of *Hansenula polymorpha* and *Candida boidinii*, which are impaired in activities of assimilatory enzymes, were isolated by a process using ethyl methanesulphonate as mutagen.[45]

An enrichment method was employed for the isolation of auxotrophic mutants of *Hansenula polymorpha*.[46] Nystatin, which is an antifungal polyene antibiotic and has the selective action against prototrophs under conditions of high metabolic activity, was introduced to the enrichment. This treatment resulted in auxotrophic isolates at a frequency of 75%.

Mutants having the altered activity of amino acid biosynthesis are derived from *Hansenula polymorpha*.[47,48]

E. Ultrastructure of Cells

1. Protoplast

Two digestive enzymes can be applied to provoke the lysis of cell walls of methanol-utilizing yeasts. The snail enzyme converts methanol-grown cells of *Candida boidinii* to spheroplasts.[49] Dziengel et al.[50] described a controlled and gentle method for preparing protoplasts of *Kloeckera* sp. No. 2201 cells by Zymolyase®, an extracellular enzyme of *Arthrobacter luteus*. In the latter case, it was found that the carbon source of the growth medium greatly influences the susceptibility of the yeast cell wall against the attack of the enzyme. Glucose-grown cells could easily be converted into protoplasts without preceding incubation with 2-mercaptoethanol, whereas methanol-grown cells required the preincubation with the thiol agent. The influence of carbon source on the structure of yeast cell walls was also shown by the experiment, which shows that the mechanical resistance to pressure and to shear forces is much greater in methanol-grown *Kloeckera* cells than in *Saccharomyces* or *Kloeckera* cells grown on glucose.

2. Microbody

A specific organelle called the microbody, which is surrounded by a single-unit membrane, was found independently in methanol-grown yeast cells by three groups of researchers.[51-53]

Fukui et al.[51] examined the relationship between catalase content and ultrastructure of methanol-utilizing yeasts and observed the occurrence of microbodies in all yeasts when they are grown on methanol and have higher catalase activity. The catalase activity detected with 3,3'-diaminobenzidine and the crystalloid structure in the microbodies were suggested to have a close relationship. Sahm et al.[52] found the microbodies in *Candida boidinii* grown on methanol and gave some evidences that the alcohol oxidase is located within these particles. They could separate the spheroplast lysates in a mitochondrial fraction and a microbody fraction by noncontinuous Ficoll® density gradients. The microbodies are 0.4 to 0.6 μm in diameter and contain crystalloid inclusions.

Gel electrophoretic studies of the microbody fraction demonstrated that it contains only two predominant protein bands consistent with alcohol oxidase and catalase.[49] No alcohol oxidase and catalase activities were found in mitochondria. Localization of the two enzymes was also ascertained in the isolated microbodies, which were prepared from the cells of *Kloeckera* sp. No. 2201 of the late exponential growth phase by Zymolyase® treatment, followed by fractionation with density-gradient centrifugation.[54] Rapid degradation of the microbodies was observed with the loss of activity of both enzymes when the methanol-grown cells were transferred to a glucose medium.[55] These observations characterize the biological function of the microbodies as peroxisomes.

The microbodies were observed as refractile granules under a phase contrast microscope and their membranes were not stained by a specific plasmalemma-staining method.[56] The surface structure was studied by scanning electron microscopy and transmission microscopy in relation to mitochondria.[57] Isolated microbodies showed a smooth surface and retained an intact structure after ion etching in contrast to mitochondria which exhibited a rough surface structure and were damaged by the ion etching. Thus, the properties of the microbody membrane and internal structure are different from those of mitochondria.

A partly or completely ordered matrix of the microbodies was found and it was expected to be an in vivo crystalline substructure of the peroxisomal contents. Osumi et al.[58-61] demonstrated a composite crystal model of the crystalloid of the microbodies of *Kloeckera* sp. No. 2201. The crystalloid shows a cubic shape with a lattice structure which is composed of two types of particles with different diameters (10 and 7 nm) in a fixed 1:1 ratio. The large and small particles are arranged alternately and the crystalloid is structurally similar to a rock salt, replacing the Cl^- and Na^+ ions in the NaCl crystal. Finally, the electron microscopic investigations using the cryosectioning technique together with electron diffraction, optical diffraction, and computer simulation led to the conclusion of three-dimensional structure of the crystalloid in which the large and small particles consist of alcohol oxidase and catalase, respectively.

On the other hand, Veenhuis et al.[62] presented another model of substructure of the crystalloid with the microbodies of *Hansenula polymorpha*. They observed that catalase was completely removed from the isolated microbodies by osmotic shock treatment. After such treatment, the ultrastructure of the crystalline matrix of the organelles remained virtually intact, and alcohol oxidase activity was still present in the matrix. Therefore, they designed a three-dimensional model in which alcohol oxidase protein is the only structural element of the peroxisomal crystalloids. The repeating unit of this structure is composed of four alcohol oxidase molecules. The crystalloid represents an open structure, which may explain the observed free mobility of catalase molecules.

3. *Array in Membrane*

Paracrystalline arrays were observed in the cell membranes of *Kloeckera* sp. No. 2201.[63] The repeating unit protruding on the surface has a distorted rectangular outline and consists of four major subunits revealing the arc-dot substructure. The unit cell is approximately identical and has equal cell size of 12 nm with an angle of about 60°. Although the biological function of the regularly arranged particles is not clear, the particle is considered to be one of the major membrane proteins. The particles exist predominantly in individual form in the membrane of the growing phase and the arrays are formed abundantly in the stationary phase.

III. METABOLIC FEATURES OF METHANOL-UTILIZING YEASTS

A. Function of Dissimilatory Enzymes

There are many unique features in the metabolic pathway of methanol-utilizing yeasts, especially in the early steps of the dissimilation and assimilation of methanol, in contrast to bacteria. The principal reaction sequence of the dissimilation involves alcohol oxidase/catalase system in peroxisomes, and formaldehyde dehydrogenase, *S*-formylglutathione hydrolase, and formate dehydrogenase in cytoplasm as shown in Figure 1. This is responsible for the complete oxidation of methanol to CO_2 through formaldehyde and formate.

1. *Alcohol Oxidase*

A unique characteristic of the methanol oxidation system of yeasts appears in the first step, methanol to formaldehyde. All yeasts examined have an alcohol oxidase (EC 1.1.3.13) which catalyzes the following reaction using molecular oxygen as the electron acceptor:

$$CH_3OH + O_2 \rightarrow HCHO + H_2O_2 \qquad (1)$$

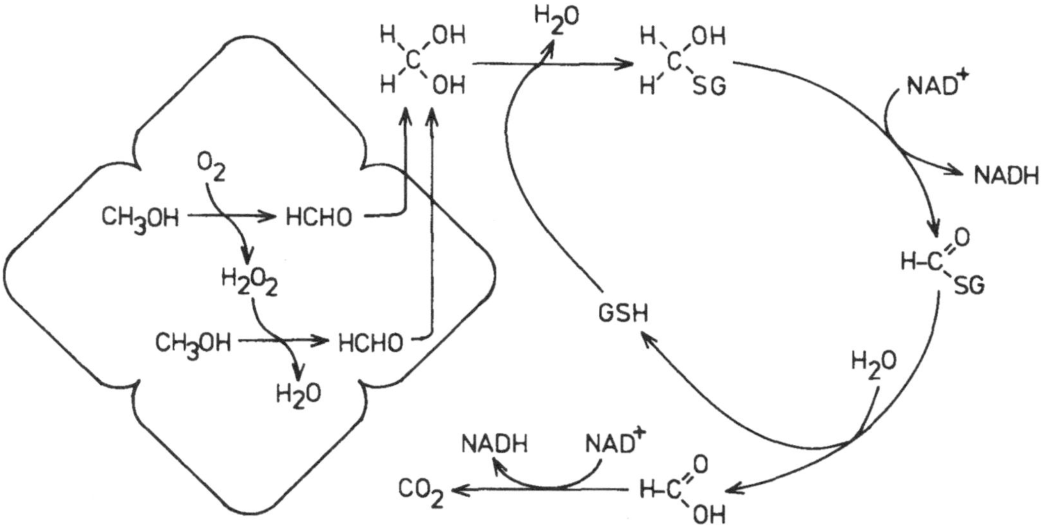

FIGURE 1. Dissimilatory sequence of methanol oxidation in methanol-utilizing yeast.

In comparison with the bacterial enzymes, this first step of methanol oxidation in yeasts is disadvantageous to ATP generation. The enzyme was first found in the cell-free extract of *Kloeckera* sp. No. 2201 and characterized by Tani et al.[64] The crystals which are obtained by a procedure involving ammonium sulfate fractionation and DEAE-cellulose and Sephadex® G-200 column chromatographies have a specific activity about 12-fold that of the crude cell-free extract. This enzyme is induced when the yeast is grown on a methanol medium and is almost equally active for ethanol and methanol; to a lesser extent it is active for several primary alcohols.[65] The oxidation occurs stereoselectively where the ratio of hydrogen abstraction in the oxidation of ethanol is about 7:1 in favor of the pro *R* hydrogen atom.[66]

The molecular weight of the enzyme is 673,000 and the enzyme is composed of eight subunits, each of which contains one coenzyme, FAD.[67] Each subunit is arranged in an octad aggregate composed to two tetragons face-to-face based on electron microscopical observation.

The purified alcohol oxidase was also obtained from cell-free extracts of *Candida* sp. N-16,[68] *Candida boidinii*,[43] *Hansenula polymorpha*,[67] *Candida* sp. 25-A,[69] and *Pichia pastoris*.[70] Physicochemical and enzymological properties of the enzyme of these yeasts are very similar to those of *Kloeckera* sp. No. 2201.

The absorption spectrum of the alcohol oxidase is unusual in flavin enzymes. The higher λ_{max} at 373 nm than that at 461 nm is explained by the presence of considerable amounts of anionic flavin radical.[71] In this case, the redox stoichiometry appears to be $2e^-$, and H_2O_2 (but not superoxide) is the first product. The absorption spectrum also indicates the presence of a chromophore other than FAD. The enzyme from *Hansenula polymorpha* is proven to contain five "red" flavin semiquinones and two oxidized flavins per octomer.[72] Addition of substrate results in the reduction of the two oxidized flavins but does not affect the flavin semiquinones.

Oxygen-dependent oxidation of formaldehyde found in the cell-free extracts of *Candida* sp. N-16[73] and *Hansenula polymorpha*[74] was evidenced to be ascribed to alcohol oxidase.[75] The enzyme possesses a dual substrate specificity towards lower primary aliphatic alcohols and formaldehyde, since loss of alcohol oxidase activity in a mutant led to loss of the formaldehyde oxidase activity. This oxidation occurs because more than 99% of formaldehyde is hydrated to form an alcoholic compound in the aqueous solution. The rather high Michaelis constant for formaldehyde, 2.40 m*M* for purified alcohol oxidase compared to that for NAD-dependent formaldehyde dehydrogenase, 0.29 m*M*, led to a conclusion that alcohol oxidase is not implicated in formaldehyde oxidation in vivo.

Alcohol oxidase is specifically induced during growth of yeasts on methanol[76] and is located in microbodies together with catalase.[54] Since formaldehyde is oxidized in the cytoplasm by NAD-dependent dehydrogenases, the rate of peroxisomal formaldehyde production is decisive for the rate of ATP synthesis by mitochondria from NADH generated in the cytoplasm. When *Hansenula polymorpha* was grown in a methanol-limited continuous culture at a dilution rate of 0.16/hr, a maximal capacity to oxidize excess methanol was 1.6 times higher than the rate required to sustain the growth rate.[39] The decrease of the dilution rate of 0.03/hr increased the capacity to 20 times. The overcapacity was explained by the content of alcohol oxidase which has poor affinity for methanol. The level of the enzyme increases from 7% to approximately 20% of the soluble protein when the growth rate is decreased from 0.16 to 0.3/hr.

The high content of alcohol oxidase, which has eight molecules of FAD in each molecule, is correlated with the content of flavin in the yeast cells. When *Kloeckera* sp. No. 2201 was grown on methanol, the amount of FAD and the activity of FAD pyrophosphorylase were three and five times higher, respectively, than those in the cells grown on glucose, ethanol, or glycerol.[77,78] A positive correlation between the formation of alcohol oxidase and FAD pyrophosphorylase was also shown with a mutant of *Candida boidinii*.[79] The mutant lacking alcohol oxidase did not inducibly form FAD pyrophosphorylase after transference of glucose-grown cells to a methanol medium, in contrast to wild-type cells.

2. *Catalase*

In methanol-utilizing yeasts, catalase must play an active role in the metabolism of methanol because the oxidation of methanol by alcohol oxidase is accompanied by the formation of a toxic compound, H_2O_2. This is supported by the fact that catalase is inducibly formed when the yeast is grown on methanol[68,76,80,81] and is located in microbodies together with alcohol oxidase.[54]

The catalase was purified from the cell-free extract of *Candida* sp. N-16 to the homogeneous state by a procedure involving ammonium sulfate fractionation and DEAE-Sephadex® and Sephadex® G-200 column chromatographies.[82] The purified enzyme has two peaks in the absorption spectrum, 279 and 407 nm, and a molecular weight of 240,000. The enzymological properties are similar to those of catalase of *Saccharomyces cerevisiae* (bakers' yeast).

Catalase also functions like peroxidase to oxidize methanol, formaldehyde, and formate in the presence of excess H_2O_2. The stimulation of formaldehyde formation from methanol by an addition of H_2O_2 to cell-free extract of methanol-grown yeast was first observed by Fujii and Tonomura[68] and then by Roggenkamp et al.[80] This seemed to be contradictory to the fact that the oxidation of methanol by alcohol oxidase is inhibited by H_2O_2. The investigations with purified enzyme preparations showed that the oxidation of methanol by alcohol oxidase is irreversibly inhibited by the addition of H_2O_2, whereas the oxidation is extremely stimulated by H_2O_2 in the presence of catalase.[83] The oxidation of methanol generates the H_2O_2. Thus, the oxidation of methanol by alcohol oxidase and catalase can be shown as follows:

$$\begin{array}{llll} CH_3OH + O_2 & \rightarrow & HCHO + H_2O_2 & \text{(alcohol oxidase)} \\ CH_3OH + H_2O_2 & \rightarrow & HCHO + 2H_2O & \text{(catalase)} \\ \hline 2CH_3OH + O_2 & \rightarrow & 2HCHO + 2H_2O & \end{array} \qquad (2)$$

Oxidation of formaldehyde and formate by the catalase was also found by van Dijken et al.[84] However, they could not conclude a possible occurrence of the further peroxidation of formaldehyde in vivo in respect to the compartmentalization of the dissimilatory system of methanol. The peroxidation of methanol has no effect on the yield of yeast biomass from methanol, whereas that of formaldehyde and formate should decrease it on the dissimilation efficiency.

3. Formaldehyde Dehydrogenase

In all species of methanol-utilizing yeasts studied so far, a glutathione-dependent dehydrogenase (EC 1.2.1.1) which uses NAD as the electron acceptor is known to catalyze formaldehyde oxidation, and the enzyme is induced when the yeast is grown on methanol.[68,85,88]

The enzyme was purified from the cell-free extract of *Kloeckera* sp. No. 2201 as well as from that of *Candida boidinii*,[87] by a procedure involving ammonium sulfate fractionation and DEAE-cellulose, hydroxyapatite, and Sephadex® G-200 column chromatographies.[85] The molecular weight of the enzyme is 80,000 and that of the subunit, 40,000. Besides formaldehyde, only methylglyoxal serves as the substrate; but no other aldehyde tested.

The dependency of formaldehyde dehydrogenase on glutathione is based on the spontaneous formation of a hemimercaptal between formaldehyde and glutathione, *S*-hydroxymethylglutathione, which is thought to be the actual substrate for the enzyme. The product of the oxidation of formaldehyde was proved to be *S*-formylglutathione by the findings that only *S*-formylglutathione but not formate was able to support the reverse reaction and that the absorbance at 240 nm increased as the result of this thioester formation during the forward reaction.[87] The oxidation of formaldehyde is shown:

$$\begin{array}{lcl} HCHO + GSH & \rightarrow & HOCH_2{-}S{-}G \\ HOCH_2{-}S{-}G + NAD^+ & \rightarrow & HCO{-}S{-}G + NADH + H^+ \\ \hline HCHO + GSH + NAD^+ & \rightarrow & HCO{-}S{-}G + NADH + H^+ \end{array} \tag{3}$$

The enzyme is located in cytoplasm and introduces the electron of formaldehyde into a respiratory chain through the reduction of NAD.

4. S-Formylglutathione Hydrolase

Although the reaction product of formaldehyde dehydrogenase, *S*-formylglutathione, can be hydrolyzed nonenzymatically to yield formate and glutathione, a hydrolyzing enzyme of the thioester was found in *Candida boidinii*.[87] The enzyme is induced when the yeast is grown on methanol and can supply formate to formate dehydrogenase.

Subsequently, a rather high Km value of formate dehydrogenase for formate and a lower one for *S*-formylglutathione have been debated. van Dijken et al.[88] suggested that *S*-formylglutathione is hydrolyzed by formate dehydrogenase followed by oxidation of enzyme-bound formate as the true substrate of the enzyme.

On the other hand, *S*-formylglutathione hydrolase (EC 3.1.2.12) was completely separated from formate dehydrogenase by DEAE-Sephacel® or 5′-AMP-Sepharose® 4B column chromatography.[89] The purified hydrolase preparation from *Kloeckera* sp. No. 2201 is electrophoretically homogeneous and contains no activity of formaldehyde and formate dehydrogenases. The enzyme has a molecular weight of 58,000 and is composed of two identical subunits. The apparent Km for *S*-formylglutathione is 0.077 m*M*. The enzyme of *Candida boidinii* was also purified to homogeneity and was shown to be very specific for *S*-formylglutathione.[90]

5. Formate Dehydrogenase

The enzyme which catalyzes the final step of methanol oxidation in yeasts is NAD-dependent formate dehydrogenase (EC 1.2.1.2).[68,86] The synthesis of the enzyme is induced by methanol and repressed by glucose.[76,86] The enzyme was purified 35- and 19-fold, respectively, from the cell-free extracts of *Kloeckera* sp. No. 2201[91] and *Candida boidinii*,[87] and also from *Pichia pastoris*.[92] These final preparations are homogeneous as judged by gel electrophoresis and sedimentation in a ultracentrifuge. The *Pichia* enzyme has a molecular weight of 94,000 and consists of two identical subunits. Antisera prepared against the *Pichia* enzyme form precipitin bands with isofunctional enzymes from other methanol-grown yeasts, but not bacteria, on immunodiffusion plates.

Km values for formate of formate dehydrogenases of methanol-utilizing yeasts are in the range of 13 to 55 m*M*. This places doubt on the physiological significance of the enzyme in an energy-giving system, although it is advantageous for the supply of formaldehyde to the assimilation pathway.

The Km value of formate dehydrogenase for *S*-formylglutathione was about 1 m*M*, determined by the assay of NAD reduction. *S*-Formylglutathione formed by formaldehyde dehydrogenase rapidly decreased upon addition of partially purified formate dehydrogenase of *Hansenula polymorpha*. The hydrolysis of *S*-formylglutathione by the enzyme was not observed in the absence of NAD. These observations led to a postulation that formate dehydrogenase hydrolyzes *S*-formylglutathione to form enzyme-bound formate, which is subsequently oxidized to CO_2.[88]

However, since formate dehydrogenase and *S*-formylglutathione hydrolase are completely separated by column chromatographies,[89,90] the hydrolytic activity of formate dehydrogenase is ascribed to a contaminating hydrolase, at least in the cases of *Kloeckera* sp. No. 2201 and *Candida boidinii*.

6. Cyclic Oxidation System of Formaldehyde

Presence of glucose 6-phosphate isomerase, glucose 6-phosphate dehydrogenase, and 6-phosphogluconate dehydrogenase in *Candida boidinii* shows that formaldehyde is possibly oxidized to CO_2 by a cyclic pathway similar to the dissimilatory ribulose monophosphate cycle in some methylotrophic bacteria. The two dehydrogenases were purified to the homogeneous state and their kinetic and regulatory properties were studied with the measurement of pool size of $NADP^+$ and NADPH in the cells.[93]

The cyclic oxidation of formaldehyde in the yeast begins with the fixation of formaldehyde to xylulose 5-phosphate, which functions intrinsically as the first step reaction of the assimilatory pathway, and is followed by the oxidation to CO_2 through the hexose monophosphate pathway. The balance of this cycle is summarized as follows:

$$HCHO + 2NADP^+ + ATP + 2H_2O \rightarrow CO_2 + 2NADPH + H^+ + ADP + Pi \qquad (4)$$

ATP is required for the sequence of reactions. These sequences should be important for the generation of NADPH as a reduction equivalent for biosynthetic purpose.

van Dijken et al.[7] calculated from the overall equation for formation of cell material from methanol that 3 to 4% of formaldehyde generated during methanol metabolism has to be oxidized in the cyclic pathway. The minor role of the cyclic oxidation system is also indicated by the facts that the levels of glucose 6-phosphate isomerase, glucose 6-phosphate dehydrogenase, and 6-phosphogluconate dehydrogenase are lower in methylotrophic growth and that mutants lacking formate dehydrogenase lose the ability to grow on methanol.[41,94]

7. Methanol Dehydrogenase

Presence of an enzyme other than alcohol oxidase to oxidize alcohols to the corresponding aldehydes in methanol-utilizing yeasts has been reported.

An NAD-dependent alcohol dehydrogenase was detected in *Candida boidinii*.[43] The enzyme, however, does not catalyze the oxidation of methanol and is constitutively formed. A mutant of the yeast, which is unable to grow on methanol as the sole source of carbon and energy, lacks alcohol oxidase but not the NAD-dependent alcohol dehydrogenase. Activity of the alcohol dehydrogenase was also observed in cell-free extracts of all methanol-utilizing yeasts tested but the formation was not induced by methanol.[76] These facts suggest that the enzyme has no physiological significance.

On the other hand, an activity of methanol-induced alcohol dehydrogenase which is linked to NAD and requires glutathione was found in *Pichia pinus* and *Candida boidinii*, and positive participation of the enzyme in methanol oxidation was assumed.[95,96] In this case, cooperative action of contaminating enzymes, such as alcohol oxidase and glutathione-dependent formaldehyde dehydrogenase, may indicate the dehydrogenase activity oxidizing methanol and reducing NAD.

8. Secondary Alcohol Dehydrogenase

Hou et al.[97] reported the oxidation of 2-butanol to 2-butanone by cell suspensions of several methanol-utilizing yeasts. The oxidation of secondary alcohols is catalyzed by cell suspensions of the yeasts grown on various C_1-compounds and also ethanol and propylamine.[98]

The enzyme, secondary alcohol dehydrogenase, was purified from *Pichia* sp.[99] as well as obligate and facultative methylotrophic bacteria.[100] The purified enzyme catalyzes the oxidation of secondary alcohols to the correponding methyl ketones in the presence of NAD. Primary alcohols are not oxidized. The enzyme has the molecular weight of 98,000 and is composed of two identical subunits.

Though the enzyme activity is detected in all strains of methanol-utilizing yeast tested, the active physiological role remains uncertain. Both enantiomorphs of 2-butanol are oxidized by the enzyme, but the (−) enantiomorph is oxidized at a higher rate.[101] This substantially differs from the oxidation with alcohol dehydrogenases from bakers' yeast and horse liver, which react strictly on the (+) isomer and equally on both enantiomorphs, respectively.

9. Formaldehyde-Reducing Enzyme

An NADH-dependent reduction of formaldehyde to methanol was found in cell-free extracts of methanol-utilizing yeasts.[102] The partially purified preparation from methanol-grown *Pichia pastoris* does not oxidize methanol to formaldehyde in the presence of NAD. However, it is active in oxidizing ethanol, propanol or butanol to their corresponding aldehydes. The substrate specificity of the reverse reaction is similar to NAD-dependent alcohol dehydrogenase of *Candida boidinii*.[43]

10. Nitrate Reduction

Excretion of ammonium was observed with some strains of *Candida boidinii* when they were grown on a methanol medium containing nitrate as the sole nitrogen source.[103] Participation of the nitrate reduction in the electron-transport system was expected, since the yeasts had an increased substrate constant for oxygen when methanol was used as the carbon source.

The excessive production of ammonium was observed under conditions of oxygen limitation. However, the share of total nitrate reduction in the electron transport of methanol-grown cultures was calculated to be about 18%. No increase in cell yield of oxygen-limited cultures due to nitrate addition was observed. Thus, it seems that nitrate reductase plays only a minor role in the dissimilation of methanol.

B. Peroxisomes for Methanol Oxidation

The microbodies which appear in methanol-grown yeast cells are characterized as peroxisomes. Biochemical characterization of the organelle shows that it contains enzymes involved in the oxidation of methanol, alcohol oxidase and catalase, and also D-amino acid oxidase and hydroxy acid oxidase.[49,54] The localization of catalase in microbodies is also demonstrated electron-microscopically by the cytochemical technique with 3, 3′-diaminobenzidine, an H_2O_2-dependent peroxidative staining.[54,104,105] This is further confirmed by an ultrathin frozen sectioning method in which the glutaraldehyde-fixed cells are incubated in 3,3-diaminobenzidine followed by osmification and then sectioned.[106] The reaction products are most clearly located in the matrix of the microbodies.

In the microbodies methanol is considered to be first oxidized to formaldehyde by alcohol oxidase and also possibly by catalase in the manner of peroxidase. Immobilized microbodies oxidize 2 mol methanol to form 2 mol formaldehyde with consumption of 1 mol O_2.[107] Addition of inhibitor of catalase reduces the formation of formaldehyde by half, without a change in the amount of O_2 consumed. This indicates a synergistic action of alcohol oxidase and catalase for the peroxisomal oxidation of methanol in living yeast cells.

When *Kloeckera* sp. No. 2201 is cultured on glucose and then transferred into a methanol medium, the microbodies of small size can be observed in 2-hr-old cells.[108] The number of microbodies per sectioned cells reaches 5 to 6 after a 4-hr cultivation. Though the number of microbodies does not change during prolonged cultivation, their size becomes larger with the passage of cultivation time. The development and ultrastructure of the microbodies of *Hansenula polymorpha* were compared in batch and chemostat cultures.[109] In batch culture, the microbodies appear as rounded or irregularly shaped organelles with crystalline nuclei. Development of the microbodies is accompanied by an increase in size of the crystalline inclusions which become completely crystalline in the stationary growth phase. The microbodies in methanol-limited chemostat are almost cubic and completely crystalline irrespective of growth rate. In both the cultures, aging of cells increases the microbodies in size and number. The change in number and volume density is related to the level of alcohol oxidase in the cells.

Disappearance of the microbodies was observed when methanol-grown cells of *Kloeckera* sp. No. 2201 were transferred to a methanol-glucose medium and cultivated for 8 hr.[81] The number of microbodies and concomitantly the activity of alcoholoxidase and catalase decrease when the cells of *Candida boidinii* precultured on methanol are transferred into an ethanol medium.[110] After a 6-hr cultivation, the microbodies are hardly detected. Also the activity of alcohol oxidase is not measurable and catalase activity is reduced to one-tenth, whereas the activities of formaldehyde dehydrogenase and formate dehydrogenase decrease only to about 70%. The rapid degradation of microbodies and decrease of alcohol oxidase and catalase activities are also observed in addition of glucose to the culture of *Hansenula polymorpha* growing on methanol. [55]

C. Regulation of Dissimilatory Enzymes

Comprehension of the regulatory mechanism in synthesis of methanol dissimilatory enzymes of methanol-utilizing yeasts should be a good tool to optimize a fermentation process to yield single-cell protein or metabolites with the yeasts.

The syntheses of alcohol oxidase, catalase, formaldehyde dehydrogenase, and formate dehydrogenase were demonstrated to be controlled by derepression-repression rather than by induction-repression.

The cell-free extract of *Hansenula polymorpha* grown on glycerol in batch culture contains up to 70% of alcohol oxidase activity compared to the extract from methanol-grown cells.[111] The dissimilatory enzymes were also found at increased specific activities during growth on sorbitol, ribose, and xylose as well as in glucose-grown cells which entered the stationary growth phase. In continuous culture, the similar results are obtained.[112] During growth under glucose limitation the dissimilatory enzymes increase with decreasing dilution rates.

The alcohol oxidases of *Hansenula polymorpha* synthesized during growth on glycerol and methanol are immunologically identical.[113] Growth on glycerol, however, is not dependent on the alcohol oxidase. The high catalase activity during growth on glycerol is also not a prerequisite for the utilization of this substrate, though it is absolutely required for growth on methanol.

Except for alcohol oxidase, the extent to which catabolite repression of the dissimilatory enzymes is relieved at low dilution rates is similar in *Hansenula polymorpha* and *Kloeckera* sp. No. 2201.[112] The level of alcohol oxidase of *Hansenula polymorpha* gradually increases with decreasing dilution rate, whereas the derepression of that of *Kloeckera* sp. No. 2201 is only observed at lower dilution rates and occurs to a much smaller extent than in *Hansenula polymorpha*. When growing in batch culture with glucose, *Hansenula polymorpha*, but not *Kloeckera* sp. No. 2201, was found to produce ethanol which might affect the synthesis of the dissimilatory enzymes.[114] Thus, it is considered that no significant difference in the regulation of these enzymes exists between *Kloeckera* sp. No. 2201 and *Hansenula polymorpha*.

The derepression of alcohol oxidase and catalase is shown to be associated with the synthesis of the microbodies of *Hansenula polymorpha* in the cells from glucose-limited chemostat culture[112] and in stationary growth phase cells grown in batch culture with glucose.[115] The occurrence of crystalline inclusions in the microbodies is strictly associated with the presence of alcohol oxidase in the organelle.

On the other hand, induction of alcohol oxidase was observed when *Hansenula polymorpha* was grown on mixed substrates.[113] The growth on ribose and methanol results in an enhanced rate of alcohol oxidase synthesis as compared to ribose alone in batch culture. Incontinuous culture, addition of methanol to glycerol-, glucose-, or sorbitol-limited cultures increases the specific activities of alcohol oxidase and other dissimilatory enzymes. The activity of alcohol oxidase is two times higher than that with methanol alone. The positive effect of methanol on the synthesis of the enzymes of the dissimilatory pathway for methanol in the presence of glucose is shown to overcome the repression of the enzyme synthesis caused by glucose.[41,42]

Results thus demonstrated so far indicate that the regulation of the dissimilatory enzymes might be under multiple controls including derepression and induction.

The enhanced utilization-rate of methanol in the culture on mixed substrates was also observed.[116] At high dilution rate, large amounts of methanol are considered to be oxidized to generate energy, since formate dehydrogenase activity increases. This is confirmed by the experiment with ^{14}C-methanol in which approximately 75% of methanol is converted to CO_2 and 25% incorporated into cell material.

Degradation of the microbodies with a concurrent loss of alcohol oxidase and catalase upon the transfer to glucose medium from methanol is another regulation system which is called the catabolite inactivation.[112] This process does not affect the cytoplasmic formaldehyde dehydrogenase.

Limited availability of the nitrogen source in medium influences the synthesis of alcohol oxidase in particular and other dissimilatory enzymes of *Kloeckera* sp. No. 2201 and *Hansenula polymorpha*.[117] As soon as nitrogen becomes a limiting substrate, the peroxisomal enzymes, alcohol oxidase and catalase, in both organisms and formaldehyde and formate dehydrogenases in *Kloeckera* sp. No. 2201 are subject to degradation. The measured rates of this catabolic inactivation show that in cells of *Hansenula polymorpha*, this process might be limited to peroxisomes; whereas in *Kloeckera* sp. No. 2201, the degradation affects cytoplasmic enzymes as well as peroxisomal ones.

Regulation of activities of the dissimilatory enzymes is also investigated. Since the dissimilatory and assimilatory pathways of methanol branch at formaldehyde, the regulatory mechanism of formaldehyde and formate dehydrogenases is important. Studies on initial velocity and product inhibition of both enzymes of *Candida boidinii* were carried out for the forward and reverse reactions.[118] ATP, ADP, and AMP inhibited the activity of both enzymes. Intracellular concentrations of NADH and ATP were in the range of the Ki values for the dehydrogenases. Therefore, the dissimilatory enzymes are considered to be regulated by these compounds. On the other hand, it is shown that formaldehyde itself influences the uptake and oxidation of methanol in two ways: by repressing the synthesis of alcohol oxidase and by noncompetitively inhibiting its activity.[119]

D. Assimilation Pathway

A pathway which includes dihydroxyacetone as a key intermediate is now established for the assimilation of methanol into cell constituents of methanol-utilizing yeasts. This pathway differs from the ribulose monophosphate pathway and the serine pathway in methlyotrophic bacteria and is called the xylulose monophosphate pathway or the dihydroxyacetone pathway. The assimilation pathway involves the fixation of formaldehyde formed from methanol to xylulose 5-phosphate, the formation of dihydroxyacteone and glyceraldehyde 3-phosphate, and the regeneration of xylulose 5-phosphate. Figure 2 illustrates the flow of the trioses to the assimilative glycolytic pathway and the regeneration cycle.

After the first isolation of methanol-utilizing yeast, there have been several attempts to determine the assimilation pathway in yeasts. However, sufficient data to explain the pathway had not been obtained. Incorporation of ^{14}C-labeled C_1-compounds to intermediate in the assimilation of methanol was first reported by Fujii and Tonomura.[120] They examined the incorporation of the compounds into *Candida* sp. N-16, by paper chromatography and radioautography.[121] The results of this primary study on the assimilation pathway suggested the presence of a route involving hexose phosphate as an early stage intermediate. Similar results were also obtained by other groups of researchers.[122,123] The special feature in the fixation of formaldehyde to ribose 5-phosphate in yeasts was that it required ATP.[124]

They proposed that the cell constituents are made by a sugar phosphate pathway similar in concept to the ribulose monophosphate pathway in methylotrophic bacteria. In this stage, the bacterial pathway was established that the true acceptor of formaldehyde is ribulose 5-phosphate rather than ribose 5-phosphate and that the reaction product is D-*arabino*-3-hexulose 6-phosphate.[125] The activity of a key enzyme of the pathway, hexulose phosphate synthase, in yeasts seemed to be inadequate to explain the growth rate of the yeasts. No phosphohexulose isomerase activity which catalyzes the isomerization of D-*arabino*-3-hexulose 6-phosphate to fructose 6-phosphate could be detected in the cell-free extracts of *Kloeckera* sp. No. 2201 and *Hansenula polymorpha*.[126]

1. Xylulose Monophosphate Pathway

van Dijken et al.[127] reinvestigated the possible occurrence of the ribulose monophosphate pathway in *Hansenula polymorpha*. The results obtained by short-term labeling experiment followed by chromatographic and radioautographic analyses showed that methanol is assimilated through hexose phosphates, but not through the ribulose monophosphate pathway. They successfullly constructed a possible route initiated by xylulose 5-phosphate-dependent fixation of formaldehyde which is mediated by transketolase and placed dihydroxyacetone as a key intermediate in methanol assimilation. Though the evidence of presence of the formaldehyde-fixing enzyme was not obtained, the induction of triokinase and fructose 1,6-*bis*phosphatase by methanol supported the proposed pathway. Babel and Loffhagen[128] also proposed a similar pathway, in which the fixation of formaldehyde is ATP-independent, but ATP is required for the generation of the glycolaldehyde donor, xylulose 5-phosphate, and glyceraldehyde 3-phosphate and ^{14}C-dihydroxyacetone formed from ^{14}C-formaldehyde.

The proposed pathway involves three key enzymes; a novel formaldehyde-fixing enzyme, dihydroxyacetone kinase, and fructose 1,6-*bis*phosphatase. The activity of the first enzyme was detected in the cell-free extract of methanol-grown *Kloeckera* sp. No. 2201 and was

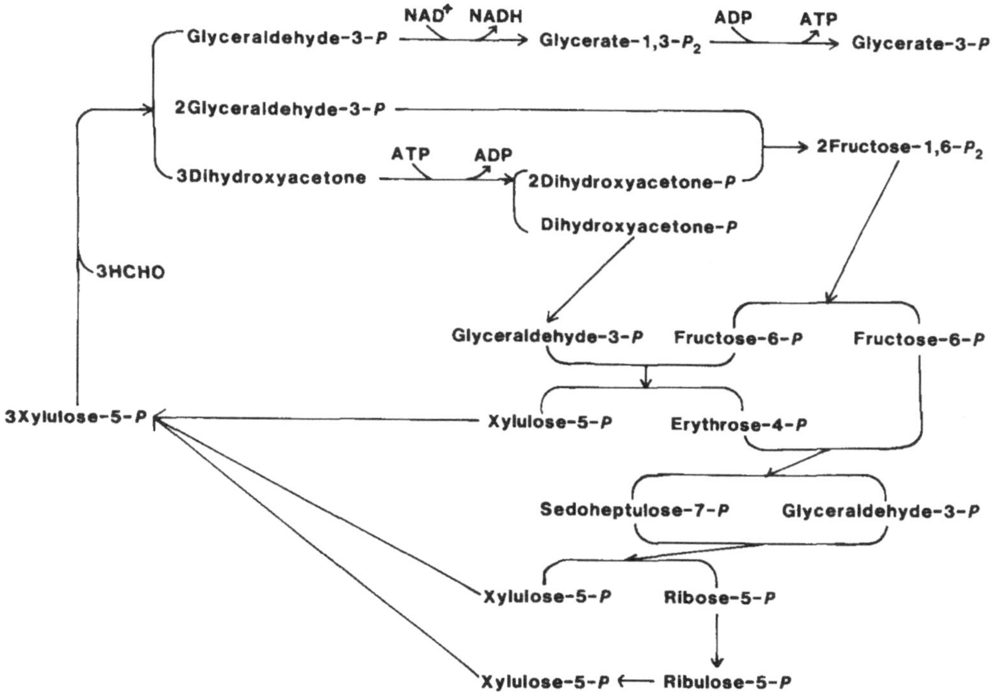

FIGURE 2. The xylulose monophosphate pathway in methanol-utilizing yeast.

partially purified.[129] No activity of the enzyme could be found in glucose-, glycerol-, or ethanol-grown cells of the yeast. ^{14}C-Dihydroxyacetone and glyceraldehyde 3-phosphate were identified as the reaction products of the fixation of ^{14}C-formaldehyde to xylulose 5-phosphate. Implication of the last two enzymes in the methanol assimilation was shown by mutant evidence.[45] Non-methanol-utilizing mutants of *Hansenula polymorpha* and *Candida boidinii* lacked the dihydroxyacetone kinase or lowered activity of fructose 1,6-*bis*phosphatase. Revertants isolated with wild-type phenotype regained the missing enzyme activities.

Comparison of the enzyme activities relating to glycolysis in cell-free extracts prepared from methanol- and glucose-grown cells of *Candida methylica* showed that methanol is assimilated by the yeast through the xylulose monophosphate pathway.[94] A rapid rise in the specific activities of dihydroxyacetone kinase, fructose 1,6-*bis*phosphatase, and dihydroxyacetone synthase was also observed during adaptation of *Candida boidinii* from glucose to methanol.[130]

Pulse-labeling experiments with whole cells of methanol-grown *Hansenula polymorpha* are also consistent with the operation of the process of formaldehyde fixation. First, dihydroxyacetone was identified as an early labeled intermediate in the assimilation of ^{14}C-methanol.[131] The labeling patterns of ^{14}C-dihydroxyacetone and ^{14}C-hexose phosphate obtained from the experiments were determined.[132] In dihydroxyacetone, over 90% of the radioactivity is located in carbon atoms 1 and 3, and in glucose, over 99% of the radioactivity is evenly distributed between carbon atoms 1, 3, 4, and 6. This is in contrast to early labeled glucose formed through the bacterial ribulose monophosphate pathway which is labeled predominantly in carbon atom 1.[7]

2. *Dihydroxyacetone Synthase*

The first step enzyme in the xylulose monophosphate pathway is a transketolase catalyzing transfer of glycolaldehyde fragment from xylulose 5-phosphate to formaldehyde as acceptor, leading to formation of glyceraldehyde 3-phosphate and dihydroxyacetone.[129,133]

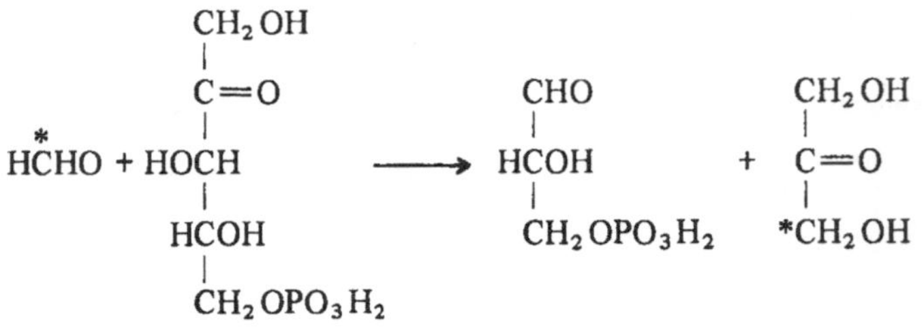

(5)

When grown on methanol *Candida boidinii* synthesizes two transketolases.[134] One is stable, exhibits only classical transketolase (EC 2.2.1.1) activity, and is found during growth both on methanol and glucose, xylose or ethanol. The other enzyme is unstable and is only present in the cells grown on methanol. The two transketolases are separated by ion-exchange column chromatography.[134,135] The former enzyme shows no appreciable activity with formaldehyde as acceptor in place of ribose 5-phosphate. The latter enzyme shows dihydroxyacetone synthase activity and also classical transketolase activity. Therefore this enzyme is characterized to be a special transketolase synthesized only during methylotrophic growth of the yeast. During adaptation from glucose to methanol, the specific activity of the dihydroxyacetone synthase rises and that of the classical transketolase falls.[130]

Dihydroxyacetone synthase was purified from the cell-free extract of *Candida boidinii*[136] and *Kloeckera* sp. No. 2201.[137] Bystrykh et al.[136] employed a purification process to last within 12 hr because of instability of the enyme and obtained the homogeneous preparation on gel-filtration and gel electrophoresis. The purified enzyme was confirmed to be different from classical transketolase in respect to substrate specificity, pH optimum and sensitivity to sulfhydryl reagents. The enzyme of *Kloeckera* sp. No. 2201 was also purified to electrophoretic homogeneity by a procedure involving Phenyl-Sepharose® CL-4B, DEAE-Sephacel®, Hypatite® C and Bio-Gel® A-1.5m column chromatographies.[137] It is characterized to be a new type of transketolase. The molecular weight of the enzyme is 190,000 and the enzyme is composed of four identical subunits. Thiamin pyrophosphate and Mg^{2+} are required for recovery of the activity of apoenzyme. With xylulose 5-phosphate as the glycolaldehyde donor, aliphatic aldehydes of C_1 to C_7, glycolaldehyde, and glyceraldehyde are better acceptors than ribose 5-phosphate.

3. *Dihydroxyacetone Kinase*

Dihydroxyacetone is phosophorylated through a specific kinase[138] and is available as dihydroxyacetone phosphate for the regeneration of the C_1 acceptor, xylulose 5-phosphate. The enzyme was purified about 100-fold from the cell-free extract of *Candida methylica* by a procedure involving streptomycin sulfate and ammonium sulfate fractionation, and Sephadex® G-200 and DEAE-cellulose column chromatographies.[139] The enzyme phosphorylates dihydroxyacetone four times faster than D,L-glyceraldehyde, and does not phosphorylate other C_3 compounds, including glycerol.

The enzyme is also involved in the metabolism of glycerol of methanol-utilizing yeasts.[140] In the yeasts which can utilize methanol and glycerol, glycerol is assimilated after oxidation to dihydroxyacetone. Thus, the dihydroxyacetone is a central intermediate in the assimilation of glycerol as well as methanol and phosphorylated by dihydroxyacetone kinase.

Dihydroxyacetone phosphate thus formed might also be metabolized through another pathway, in which it is converted to methylglyoxal by methylglyoxal synthase.[141] The enzyme is constitutively formed in methanol-utilizing yeasts and assumed to be responsible for additional conversion of dihydroxyacetone phosphate or glyceraldehyde 3-phosphate to pyruvate through lactate as a bypass. Methylglyoxal, which has been known as a lethal factor, is possibly involved in the methanol metabolism not only as the assimilatory bypass but also as the only substrate besides formaldehyde for formaldehyde dehydrogenase of *Kloeckera* sp. No. 2201[85] and *Candida boidinii.*[87]

E. Energetics

Pathways to dissimilate and assimilate methanol are summarized in terms of cell yield where ATP is the principal compound. The oxidation of methanol to CO_2 functions as an ATP-producing system. The pathway leading to the cell material described as $C_4H_8O_2N$, from the C_1-compound through 3-phosphoglycerate, a common intermediate in the biosynthesis of the cell material, is represented to be an ATP-consuming system. This scheme makes it possible to calculate a theoretical growth yield of organisms grown on methanol.[74]

The energetics of methanol-utilizing yeasts should be highly influenced by the efficiency of oxidative phosphorylation system. However, only limited studies have been done to obtain more information about this problem in methanol-grown cells. A preliminary test of the respiratory characteristics was shown with a preparation of intact and well-respiring mitochondria of the yeast.[142] It is suggested that the terminal respiration is partly mediated by cytochrome oxidase since the respiration on methanol is inhibited by 50 to 80% by 0.2 m*M* KCN, a potent inhibitor of cytochrome oxidase-mediated respiration.[38]

van Dijken et al. have discussed the mitochondrial ATP synthesis from NADH in methylotrophic growth of the yeast. Since formaldehyde and formate dehydrogenases reducing NAD^+ on methanol oxidation are located in the cytoplasm of yeast cells, most of the NADH required for the generation of ATP might be generated outside the mitochondria. This contrasts the situation during growth on other carbon substrates where most or all of the NADH is generated inside the mitochondria by pyruvate dehydrogenase and the TCA cycle enzymes. The inner membrane of mitochondria is impermeable to NADH. They suggest that NADH dehydrogenase localized on the outer surface of the inner membrane oxidizes the cytoplasmic NADH. Since this pathway of oxidation does not associate with phosphorylation at site I, NADH from methanol oxidation probably generates no more than 2 mol of ATP.

The significance of the TCA cycle in energy metabolism is considered to be quite limited in methylotrophic growth.[94] The activities of most of the enzyme of the cycle in methanol-grown *Candida methylica* are lower than those in glucose-grown cells. The importance of the role of the TCA cycle as the biosynthetic process in methylotrophic growth is, however, emphasized by the facts that an intensive fixation of CO_2 occurs in the presence of methanol with the primary formation of aspartate and malate, and pyruvate carboxylase is much more active in methanol-grown cells than in glucose-grown cells.

ATP balance of the assimilation pathway of methanol can be roughly calculated in comparison between yeasts and bacteria in spite of the variation of the dissimilation and assimilation pathways and the efficiency of oxidative phosphorylation system. The serine pathway consumes three molecules of ATP and one molecule of NADH and reduces one molecule of flavoprotein in the formation of one molecule of 3-phosphoglycerate. In the formation of the compound, the ribulose monophosphate pathway gives one molecule of NADH. The xylulose monophosphate pathway in methanol-utilizing yeasts consumes three molecules of ATP to form one molecule of glyceraldehyde 3-phosphate in phosphorylation of dihydroxyacetone, regenerating three molecules of xylulose 5-phosphate. The consumed ATP can be compensated in the oxidation of glyceraldehyde 3-phosphate to 3-phosphoglycerate yielding one molecule each of ATP and NADH.

F. Applied Aspects

1. Single-Cell Protein

The first aim of utilization of methanol by microorganisms especially by yeasts was to produce single-cell protein. Current knowledge of methanol metabolism may be sufficient to produce a theoretical cell yield value, although various factors besides the metabolic pathway must be considered to obtain the actual yield data. van Dijken and Harder[74] reported the yield of microorganisms grown on methanol, based on a value of Y_{ATP} of 10.5. The calculation showed a lower cell yield of methanol-utilizing yeasts (0.54 g/g methanol) than for methanol-utilizing bacteria (0.63 g/g methanol). They further obtained a value of the low yield for *Hansenula polymorpha* grown in a methanol-limited chemostat, 0.38 g/g methanol, which was due to H_2O_2-producing oxidation of methanol by alcohol oxidase.[39]

There have been several examples on the growth of methanol-utilizing yeasts. Early estimation of the yeast as single-cell protein was made by Cooney and Levin.[143] A thermotolerant yeast isolated by them, *Hansenula polymorpha,* had a maximal specific growth rate of 0.22/hr at 37 to 42°C, a maximal growth yield of 0.36 g/g of methanol and 0.37 g/g of O_2 and a highest cell density of 24.4 g/ℓ in continuous culture with 0.4% methanol medium.[15] Maximal cell yield and specific growth rate of a thermotolerant *Candida boidinii* were 49.5% (w/w) and 0.2/hr, respectively.[28] The productivity of *Pichia methanothermo* reached 4.12 g of cells per liter per hour in continuous culture in medium containing 73 g of ethanol per liter, with a dilution rate of 0.146/hr, temperature of 40°C, and cell density of 28.2 g/ℓ.[18] The maximum cell yield in the continuous culture was 0.386 g/g of methanol.

When yeast is used as feed or food supplement, it is desired for the yeast cell to have not only high cell yield and growth rate but also high protein content and low nucleic acid content. The crude protein contents were 45.3% for *Kloeckera* sp. No. 2201,[11] 46% for *Hansenula polymorpha,*[15] 47.3 to 53.8% for *Pichia methanothermo,*[18] 45.7% for a thermotolerant *Candida boidinii,*[28] and 59.5% for *Candida methanophilum.*[29] The nucleic acid contents of these yeasts were 5.4%,[11] 5 to 7%,[15] 3.4%,[28] and 2.82 to 3.36% for *Torulopsis methanosorbosa.*[31]

Fermentor design has been improved and allows the use of methanol for the industrial process of yeast cell production. An air-lift fermentor was made as a pilot plant fermentor to obtain high oxygen transfer and miscibility.[144] A high cell density culture of *Pichia aganobii* was carried out at a normal dilution rate.[145]

2. Amino Acid Production

Only L-tryptophan excretion by mutants of *Hansenula polymorpha* was reported for the amino acid by methanol-utilizing yeasts. A bradytrophic mutant, which required the supplement of both L-tyrosine and L-phenylalanine for the growth and grew later on both unsupplemented and supplemented media, excreted a low but significant level of L-tryptophan, 28 mg/ℓ, when grown in a medium containing 2% methanol as the major carbon and energy source.[47]

Hansenula polymorpha showed greater sensitivity to DL-5-fluorotryptophan, an antimetabolite of tryptophan, when growing on methanol than when growing on glucose. The mutants resistant to the antimetabolite enhanced the accumulation of indoles including tryptophan and/or tryptophan metabolites.[48] The overproduction was due primarily to a partial desensitization of anthranilate synthase to feedback inhibition by L-tryptophan.[146]

3. Citric Acid Production

Candida lipolytica and many other strains of the genus *Candida* are known to accumulate extracellularly citric acid from *n*-paraffins and other carbon compounds. Mutant strains of these yeasts, which are sensitive to fluoroacetate, a potent inhibitor of aconitase, have lower activity of aconitase and high citric acid productivity.

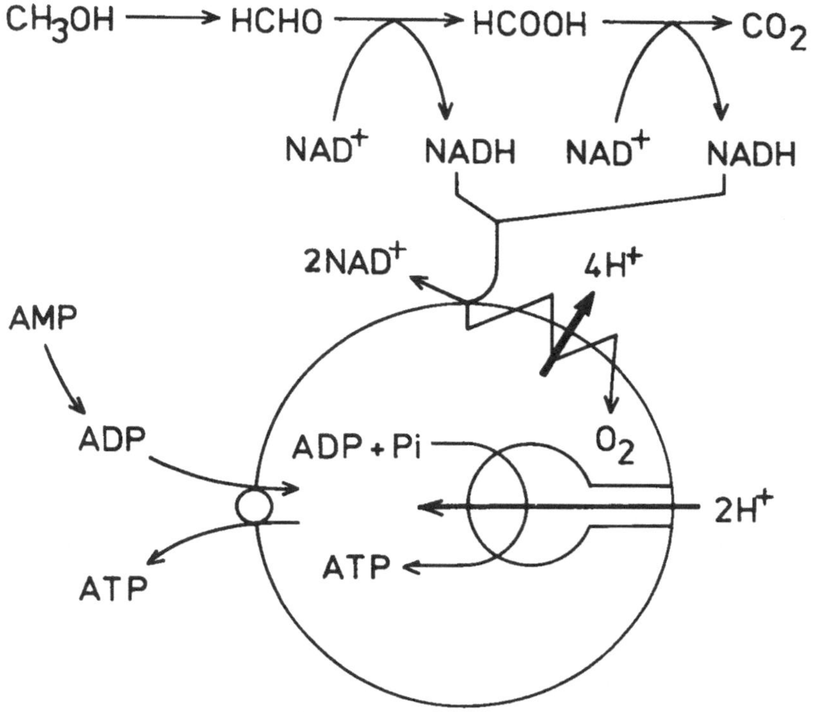

FIGURE 3. Reaction sequences of ATP production with *Kloeckera* sp. No. 2201.

Excretion of citric acid of methanol-utilizing yeast was described for *Candida boidinii*, which produced about 1 g of citric acid per liter when fluoroacetate was added to a methanol medium.[4]

On the contrary, it was found that mutants of *Candida* sp. Y-1, which were resistant to fluoracetate, produced a considerable amount of citric acid in a methanol medium.[147] About 5 g of citric acid per liter of the medium was obtained after the 4-day cultivation of a mutant.

4. Nucleotide Coenzyme Production

Methanol-utilizing yeasts possess high content of alcohol oxidase which consists of eight molecules of FAD in each enzyme molecule. The increase of FAD content in the cell and the derepression of FAD pyrophosphorylase, the last step enzyme in FAD biosynthesis, have been observed on methanol-grown yeasts.[77-79] The induction of FAD biosynthesis by methanol led to the investigation of the production of FAD by methanol-utilizing yeasts. Riboflavin or FMN added to a yeast culture on a methanol medium was converted to FAD at a good yield, 45.4 mg/ℓ.[148]

A novel process of ATP production was developed using the cells of *Kloeckera* sp. No. 2201. Methanol-utilizing yeasts reduce NAD^+ in the oxidation pathway to supply energy for the assimilation of C_1-compounds. The NADH formed should be led to the respiratory chain to phosphorylate ADP. Thus, the chemical energy of C_1-compounds can be delivered to ATP in the yeast. Based on this principle, Tani et al.[149] constructed a reaction mixture to produce ATP from AMP and C_1-compound. The process is a result of sequential reactions including the oxidative phosphorylation system together with the oxidation of methanol or formate as the C_1-substrate and the phosphorylation of AMP to ADP by adenylate kinase as shown in Figure 3.

The NADH-yielding enzymes in methanol oxidation, formaldehyde dehydrogenase and formate dehydrogenase, are located in cytoplasm, whereas the oxidative phosphorylation system and a large part of the adenylate kinase are located in mitochondria. The preparation of cells which are treated by Zymolyase® or by the addition of sorbitol into cell suspension at 37°C retains these activities and is permeable for the phosphorylated substrate and product. The maximum yield of ATP is 30 g/ℓ in 60 to 70% conversion rate against AMP. This process is characterized as a system for energy conversion of reduced C_1-compounds to a useful compound.

5. Oxidized Metabolite Production

A process to produce formaldehyde from methanol with *Hasenula polymorpha* was presented.[150] The high activities of alcohol oxidase and catalase obtained when the yeast was grown on methanol in continuous culture at a low dilution rate made this high-yield process possible. A method for preparing the cell-free extract under the conditions defined for maximum release and retention of the high activities was developed. Partially purified enzyme system in the free and immobilized forms could completely oxidize methanol at a concentration as high as 100 m*M* in a batch reactor. Subsequently, immobilized whole cells of the yeast was applied to the oxidation process.[151] The reaction with immobilized whole cells which omitted the enzyme preparation process, was not inhibited by high methanol concentration, though the conversion yield was lower than that with immobilized enzyme. The process of enzymatic oxidation of methanol has several advantages over the catalytic chemical process in less energy consumption and in higher specificity.

Several methyl ketones are prepared with oxidation of their corresponding secondary alcohols by secondary alcohol dehydrogenase of methanol-utilizing yeasts. Cell suspensions of the yeasts have the higher production rate and the higher temperature optimum for 2-butanone production from 2-butanol than those of methylotrophic bacteria.[97] *Candida utilis, Hansenula polymorpha, Pichia* sp., *Torulopsis* sp., and *Kloeckera* sp. grown on methanol, methylamine, methylformate, or propylamine oxidized isopropanol, 2-butanol, 2-pentanol, and 2-hexanol to acetone, 2-butanone, 2-pentanone, and 2-hexanone, respectively. Among them, 2-butanol was oxidized at the highest rate, at 6.9 μmol/hr/mg of protein.[98]

6. Immobilization of Microbody

Microbodies isolated from methanol-grown cells of *Kloeckera* sp. No. 2201 were immobolized by two types of entrapping techniques: photo-crosslinking of liquid resin oligomers and crosslinking of albumin molecules with glutaraldehyde.[107,152] The immobilized microbodies showed activities of alcohol oxidase, catalase, and D-amino acid oxidase at 70 to 80, 40 to 50 and 50% of those in the free microbodies, respectively. Enzymatic properties of the immobilized microbodies were the same as those of the free ones. The preparation could be useful as a multifunctional biocatalyst for analytical purposes and for production of useful metabolities.

IV. OTHER METHYLOTROPHIC EUKARYOTES

A. Methylated Amine-Utilizing Yeasts and Molds

Growth of yeast on methylated amine as the sole carbon source was reported by Yamada et al.[153] *Candida* sp. WY-3, which was initially isolated for the oxidation of *N*-methyl-*n*-butylamine to *n*-butylamine, was able to grow on a number of tertiary, secondary, and primary amines as the sole nitrogen source. The yeast showed low but significant growth on the medium containing 0.1% methylamine as well as other C_1-compounds, methanol, formaldehyde, and formate, as the sole carbon source. Although the growth on methylamine as the sole carbon source has not been further studied, the yeast can be recognized as facultative methylotroph.

Utilization of methylamine as the carbon source by methanol-utilizing yeasts was described for *Candida utilis, Hansenula polymorpha, Pichia* sp., *Torulopsis* sp., and *Kloeckera* sp., though the medium used contained yeast extract.[98]

van Dijken and Boss,[154] however, could not find any yeast strain able to utilize primary and methylated amines as the sole source of carbon and energy in a screening of 461 yeast strains. The amines used in the screening were methylamine, dimethylamine, trimethylamine, ethylamine, propylamine, butylamine and benzylamine, tetramethylammonium chloride, and choline. The majority of the yeasts (86%) were able to utilize one or more of these amines as nitrogen source.

When yeasts were grown on a medium containing methylamine as nitrogen source, large microbodies appeared in cells of both the methylotrophic yeasts, *Hansenula polymorpha*, and the nonmethylotrophic yeast, *Candida utilis.*[155] Synthesis of enzymes involved in the methylamine oxidation paralleled with the development of microbody. Results obtained from cytochemical staining experiments indicated that the amine oxidase activity was located in the microbodies in both organisms. Catalase activity was also found in the organelles, which could therefore be recognized as peroxisomes.

The microbodies contained both activities of alcohol oxidase and amine oxidase when *Hansenula polymorpha* was grown on a medium containing methanol and methylamine as the sole carbon and nitrogen sources, respectively.[156] A completely crystalline substructure of the microbodies was observed in the methanol-methylamine-grown cells. Transfer of cells grown on methanol-methylamine medium into glucose-ammonium sulfate medium resulted in the loss of activities of both alcohol oxidase and amine oxidase. This is due to the degradation of the original crystalline microbodies. Immediately after the cells were transferred into glucose-methylamine medium, the activities of both alcohol oxidase and amine oxidase decreased. However, only amine oxidase activity increased again after 2 hr of incubation. This subsequent rise in amine oxidase activity was associated with the development of new microbodies in the cells in which degradation of original microbodies continued.

On the other hand, the utilization of a tertiary methylated amine, choline, as the sole source of carbon and nitrogen was found in molds. Strains belonging to the genera *Cylindrocarpon, Penicillium, Aspergillus, Fusarium* and *Gibberella* might metabolize choline through demethylation, resulting in the formation of formaldehyde.[157] Among them, the dissimilatory metabolism by *Cylindrocarpon didymum* has been investigated in detail: it includes oxidation of choline to betaine aldehyde by choline oxidase,[158,159] oxidation of betaine aldehyde to betaine by NAD-dependent betaine aldehyde dehydrogenase,[160] demethylation of betaine to dimethylglycine, oxidative demethylation of dimethylglycine to sarcosine,[161] and oxidative demethylation of sarcosine to glycine.[162] In this dissimilatory sequence, three new oxidases, which were O_2-dependent and H_2O_2-producing, were found responsible for the oxidation of choline, dimethylglycine, and sarcosine, respectively. The last two steps of choline oxidation produce one molecule each of formaldehyde which might be metabolized as the source of carbon and energy by C_1-metabolism.

B. Methanol-Utilizing Molds

Three strains of mold growing on methanol as the sole source of carbon and energy have been known and characterized as facultative methylotroph. From the technological view on single-cell protein production, mold in general has some advantages such as easy harvesting and low content of nucleic acid.

Trichoderma lignorum was isolated from soil surrounding petroleum storage tanks as methanol-utilizer.[163] The mold can also grow on methylamine and formate but not formaldehyde. Final cell yield in a batch culture reached 0.31 g cells per liter at the expense of 3.7 g of methanol per liter. Growth inhibition was observed with methanol concentrations in excess of 4.7 g/ℓ.[164] The optimum temperature for the cell growth was 23°C.

Two molds were isolated from soil which grew on 0.1 to 0.2% formaldehyde as the sole carbon source and were identified as *Gliocladium deliquescens* and *Paecilomyces varioti*.[165] The yield of mycelia was 0.24 to 0.32 g/ℓ with 86 to 93% consumption of formaldehyde in the case of 0.1% formaldehyde medium. Both the strains can also grow on a medium containing 5% methanol or 5% Na-formate. The former strain can grow on higher methanol concentration of 7%.

Cell-free extract of methanol-grown *Paecilomyces varioti* oxidizes methanol and formaldehyde by NAD-dependent dehydrogenation. The extract of *Gliocladium deliquescens,* to the contrary, has no activity of NAD- or NADP-dependent methanol oxidation. However, it has activity of methanol-dependent methylene blue reduction. The extract of ethanol-grown cells does not possess this activity. The oxidation of formaldehyde by the *Gliocladium* extract is dependent on NAD. The assimilation pathway of the two molds is considered to be the serine pathway since serine and malate appear in the early time of ^{14}C-C_1-compound incorporation experiments and the activities of hydroxypyruvate reductase and phosphoenolpyruvate carboxylase are much higher in methanol-grown cells than in ethanol-grown cells.

C. Methane-Utilizing Yeasts

Growth on methane has been restricted to prokaryotic organisms. Wolf and Hanson[166] first isolated yeast strains utilizing methane as the sole source of energy under an atmosphere of 70% methane/20% air/10% CO_2. In 10% CO_2 atmosphere, cells incorporated little or no labeled carbon from ^{14}C-CH_4 into cellular materials. However, in the atmosphere without added CO_2, a large amount of radioactivity from ^{14}C-CH_4 was incorporated into the cells. The highest cell yield and the fastest generation time were 2.1 g/ℓ and 2 days, respectively. The yeasts were identified as *Sporobolomyces roseus, Sporobolomyces gracilis, Rhodotorula glutinis*, and *Rhodotorula rubra*.[167] All were characterized as facultative methylotrophs, with no ability to grow on methanol or methanol vapor. All the type strains of yeasts tested failed to utilize methane.

Microbodies were detected in cells of *Sporobolomyces roseus* and *Rhodotorula glutinis* when grown on methane but rarely when grown on glucose.[168] The size and number of the microbodies in cells did not increase with culture age and no crystalline organization was observed within the organelle. These are distinct from those of methanol-utilizing yeasts.[52,108] Catalase activity was detected in the microbodies of hexadecane-grown cells of the methane-utilizing *Rhodotorula glutinis* by a cytochemical staining using 3,3′-diaminobenzidine. It seems that the microbodies have a significant role in methane metabolism as in methanol oxidation of methanol-utilizing yeasts.

D. Oxidation of Methanol and Formaldehyde by Nonmethylotrophic Mold and Yeasts

Alcohol oxidase of a lignin-degrading brown rot fungus, *Poria contigua*, which was very active both in the demethylation of lignin and in the oxidation of methoxyl groups to CO_2, was purified.[169] The purified enzyme is homogeneous as judged by gel electrophoresis and sedimentation in a ultracentrifuge. The molecular weight of the enzyme is estimated approximately 610,000 and the enzyme is composed of eight identical subunits, each of which contains one coenzyme, FAD. The enzyme oxidizes not only methanol but also lower primary alcohols, 2-propin-l-ol and formaldehyde. Although these enzymatic properties are very similar to alcohol oxidase of methanol-utilizing yeasts, the *Poria* enzyme is immunologically different from that of *Candida boidinii*.

A black yeast, *Aureobasidium pullulans,* which was isolated as phenol- and *o*-cresol-decomposer, showed consumption of a mixture of methanol and formaldehyde in a phenol medium. The consumption of methanol proceeded significantly in concentrations of 3 and 4% but not in 1%, and that of formaldehyde in 0.005 to 0.012%. Growth of the yeast increased with further addition of methanol and formaldehyde up to 4% and 0.01%, respectively. No detailed information is available for the methylotrophic growth and metabolism of the phenol-assimilating yeast.[170]

Debaryomyces vanriji and *Trichosporon penicillatum*, which were isolated as formaldehyde-resistant yeasts, consumed formaldehyde from 0.15 to 0.55% in a medium containing glucose as the carbon source.[171] The yeasts grew in the medium after the consumption of formaldehyde. Significantly high activities of formaldehyde dehydrogenase, *S*-formylglutathione hydrolase and formate dehydrogenase were detected in the cell-free extracts of the yeasts. Therefore, the resistance of the yeasts to formaldehyde is ascribed to detoxication by the oxidation sequence of formaldehyde.

REFERENCES

1. **Ogata, K., Nishikawa, H., and Ohsugi, M.,** A yeast capable of utilizing methanol, *Agric. Biol. Chem.*, 33, 1519, 1969.
2. **Wickerham, L. J. and Burton, K. A.,** Carbon assimilation tests for the classification of yeasts, *J. Bacteriol.*, 56, 363, 1948.
3. **Cooney, C. L. and Levine, D. W.,** Microbial utilization of methanol, *Adv. Appl. Microbiol.*, 15, 337, 1972.
4. **Sahm, H.,** Metabolism of methanol by yeasts, *Adv. Biochem. Eng.*, 6, 77, 1977.
5. **Tani, Y., Kato, N., and Yamada, H.,** Utilization of methanol by yeasts, *Adv. Appl. Microbiol.*, 24, 165, 1978.
6. **Tani, Y. and Yamada, H.,** Microbial utilizatin of C_1-compounds, *Biotechnol. Bioeng.*, 22, (Suppl. 1), 163, 1980.
7. **van Dijken, J. P., Harder, W., and Quayle, J. R.,** Energy transduction and carbon assimilation in methylotrophic yeasts, in *Microbial Growth on C_1 Compounds*, Dalton, H., Ed., Heyden and Son, London, 1981, 191.
8. **Aroj, K. A., Prabhu, K. A., and Shukla, J. P.,** Studies on the utilization of methanol by yeasts, Presented at 36th Convention Sugar Technologists Association of India, 1968.
9. **Ogata, K., Nishikawa, H., Ohsugi, M., and Tochikura, T.,** Studies on the production of yeasts. I. A yeast utilizing methanol as a sole carbon source, *J. Ferment. Technol.*, 48, 389, 1970.
10. **Lee, J. D. and Komagata, K.,** Taxonomic study on methanol-assimilating yeasts, *J. Gen. Appl. Microbiol.*, 26, 133, 1980.
11. **Ogata, K., Nishikawa, H., Ohsugi, M., and Tochikura, T.,** Studies on the production of yeast. II. The cultural conditions of methanol assimilating yeast, *Kloeckera* sp. No. 2201, *J. Ferment. Technol.*, 48, 470, 1970.
12. **Hazeu, W., de Bruyn, J. C., and Bos, P.,** Methanol assimilation by yeasts, *Arch. Mikrobiol.*, 87, 185, 1972.
13. **Ogata, K., Tani, Y., and Kato, N.,** Oxidation of methanol by yeasts, in *Microbial Growth on C_1-Compounds*, The Organizing Committee, Ed., Maruzen, Tokyo, 1975, 99.
14. **van Dijken, J. P. and Harder, W.,** Optimal conditions for the enrichment and isolation of methanol-assimilating yeasts, *J. Gen. Microbiol.*, 84, 409, 1974.
15. **Levine, D. W. and Cooney, C. L.,** Isolation and characterization of a thermotolerant methanol-utilizing yeasts, *Appl. Microbiol.*, 26, 982, 1973.
16. **Minami, K., Yamamura, M., Shimizu, S., Ogawa, K., and Sekine, N.,** Screening method for thermophilic and high growth rate ethanol yeasts, *J. Gen. Appl. Microbiol.*, 24, 155, 1978.
17. **Minami, K., Yamamura, M., Shimizu, S., Ogawa, K., and Sekine, N.,** A new fermentor with a high oxygen transfer capacity, *J. Ferment. Technol.*, 56, 64, 1978.
18. **Minami, K., Yamamura, M., Shimizu, S., Ogawa, K., and Sekine, N.,** A new methanol-assimilating, high productive, thermophilic yeast, *J. Ferment. Technol.*, 56, 1, 1978.
19. **Pal, H. S. and Hamdan, I. Y.,** Growth of a methanol-utilizing yeast, *Enzyme Microbiol. Technol.*, 1, 265, 1979.
20. **Volfová, O. and Pilát, P.,** Studies on methanol-oxidizing yeasts. I. Isolation and growth studies, *Folia Microbiol.*, 19, 249, 1974.
21. **Tonomura, K., Kanzaki, F., and Kambayashi, A.,** Isolation and characterization of methanol-assimilating yeasts, *Rep. Ferment. Res. Inst. Chiba, Jpn.*, 41, 49, 1972.

22. **Kato, K., Kurimura, Y., Makiguchi, N., and Asai, Y.**, Determination of methanol strongly assimilating yeasts, *J. Gen. Appl. Microbiol.*, 20, 123, 1974.
23. **Henninger, W. and Windisch, S.**, *Pichia lindnerii* sp. n., a new mtehanol assimilating yeast from soil, *Arch. Microbiol*, 105, 47, 1975.
24. **Lee, J. D. and Komagata, K.**, *Pichia cellobiosa, Candida cariosilignicola,* and *Candida succiphila,* new species of methanol-assimilating yeasts, *Int. J. Syst. Bacteriol.*, 30, 514, 1980.
25. **Asai, Y., Makiguchi, N., Shimada, M., and Kurimura, Y.**, New species of methanol-assimilating yeasts, *J. Gen. Appl. Microbiol.*, 22, 197, 1976.
26. **Oki, T., Kouno, K., Kitai, A., and Ozaki, A.**, New yeasts capable of assimilating methanol, *J. Gen. Appl. Microbiol.*, 18, 295, 1972.
27. **Sahm, H. and Wagner, F.**, Microbial assimilation of methanol. Isolation and characterization of the yeast *Candida boidinii, Arch. Microbiol.*, 84, 29, 1972.
28. **Tezuka, H., Nakahara, T., Minoda, Y., and Yamada, K.**, Production of yeast cells from methanol, *Agric. Biol. Chem.*, 39, 285, 1975.
29. **Mimura, A., Wada, M., Nakano, T., Hayakawa, S., and Iguchi, T.**, Isolation and characterization of a methanol-assimilating yeast, *J. Ferment. Technol.*, 56, 443, 1978.
30. **Asthana, H., Humphrey, A. E., and Moritz, V.**, Growth of yeast on methanol as the sole carbon substrate, *Biotechnol. Bioeng.*, 13, 926, 1971.
31. **Yokote, Y., Sugimoto, M., and Abe, S.**, Yeasts utilizing methanol as a sole carbon source, *J. Ferment. Technol.*, 52, 201, 1974.
32. **Pilát, P. and Prokop, A.**, The effect of methanol, formaldehyde, and formic acid on growth of *Candida boidinii* 11Bh, *Biotechnol. Bioeng.*, 17, 1717, 1975.
33. **Goto, S., Okamoto, R., Kuwajima, T., and Takamatsu, A.**, Growth characteristics of methanol-assimilating yeasts on various substrates, *J. Ferment. Technol.*, 54, 213, 1976.
34. **Reuss, M., Gnieser, J., Reng, H. G., and Wagner, F.**, Extended culture of *Candida boidinii* on methanol, *Eur. J. Appl. Microbiol.*, 1, 295, 1975.
35. **Minami, K., Yamamura, M., and Shimizu, S.**, Method and apparatus for automatic drawing of growth curve, *J. Ferment. Technol.*, 56, 229, 1978.
36. **Held, W., Schlanderer, G., Reimann, J., and Dellweg, H.**, Continuous culture of *Candida boidinii* variant 60 growing on methanol, *Eur. J. Appl. Microbiol. Biotechnol.*, 6, 127, 1978.
37. **Minami, K., Yamamura, M., Shimizu, S., Ogawa, K., and Sekine, N.**, Methods and apparatus for methanol feeding with the less growth-inhibition, *J. Ferment. Technol.*, 56, 35, 1978.
38. **Middelhoven, W. J., Berends, J., van Aert, A. J. M., and Bruinsma, J.**, The substrate constant for dissolved molecular oxygen of methanol-assimilating yeasts, *J. Gen. Microbiol.*, 93, 185, 1976.
39. **van Dijken, J. P., Otto, R., and Harder, W.**, Growth of *Hansenula polymorpha* in a methanol-limited chemostat. Physiological responses due to the involvement of methanol oxidase as a key enzyme in methanol metabolism, *Arch. Microbiol.*, 111, 137, 1976.
40. **Schlanderer, G., Held, W., Reimann, J., and Dellweg, H.**, Metabolic changes during substrate shifts in continuous cultures of the yeast *Candida boidinii* variant 60, *Eur. J. Appl. Microbiol. Biotechnol.*, 6, 133, 1978.
41. **Egli, T., Käppeli, O., and Fichter, A.**, Regulatory flexibility of methylotrophic yeasts in chemostat cultures: simultaneous assimilation of glucose and methanol at a fixed dilution rate, *Arch. Microbiol.*, 131, 1, 1982.
42. **Egli, T., Käppeli, O., and Fichter, A.**, Mixed, substrate growth of methylotrophic yeasts in chemostat culture: influence of the dilution rate on the utilization of a mixture of glucose and methanol, *Arch. Microbiol.*, 131, 8, 1982.
43. **Sahm, H. and Wagner, F.**, Microbial assimilation of methanol. The ethanol- and methanol-oxidizing enzymes of the yeast, *Candida boidinii, Eur. J. Biochem.*, 36, 250, 1973.
44. **Eggeling, L. and Sahm, H.**, Direct enzymatic assay for alcohol oxidase, alcohol dehydrogenase, and formaldehyde dehydrogenase in colonies of *Hansenula polymorpha, Appl. Environ. Microbiol.*, 39, 268, 1980.
45. **O'Connor, M. L. and Quayle, J. R.**, Mutants of *Hansenula polymorpha* and *Candida boidinii* impaired in their ability to grow on methanol, *J. Gen. Microbiol.*, 11, 203, 1979.
46. **Sanchez, S. and Demain, A. L.**, Enrichment of auxotrophic mutants in *Hansenula polymorpha, Eur. J. Appl. Microbiol.*, 4, 45, 1977.
47. **Sanchez, S. and Demain, A. L.**, Tryptophan excretion by a bradytroph of *Hansenula polymorpha* growing in methanol, *Appl. Environ. Microbiol.*, 35, 459, 1978.
48. **Denenu, E. O. and Demain, A. L.**, Derivation of aromatic amino acid mutants from a methanol-utilizing yeast, *Hansenula polymorpha, Appl. Environ. Microbiol.*, 41, 1088, 1981.
49. **Roggenkamp, R., Sahm, H., Hinkelmann, W., and Wagner, F.**, Alcohol oxidase and catalase in peroxisomes of methanol-grown *Candida boidinii, Eur. J. Biochem.*, 59, 231, 1975.

50. **Dziengel, A., Held, W., Schlanderer, G., and Dellweg, H.,** Controlled preparation of protoplasts of *Kloeckera* sp. 2201 grown on methanol, *Eur. J. Appl. Microbiol.*, 4, 21, 1977.
51. **Fukui, S., Tanaka, A., Kawamoto, S., Yasuhara, S., Teranishi, Y., and Osumi, M.,** Ultrastructure of methanol-utilizing yeast cells: appearance of microbodies in relation to high catalase activity, *J. Bacteriol.*, 123, 317, 1975.
52. **Sahm, H., Roggenkamp, R., Wagner, F., and Hinkelmann, W.,** Microbodies in methanol-grown *Candida boidinii, J. Gen. Microbiol.*, 88, 218, 1975.
53. **van Dijken, J. P., Veenhuis, M., Kreger-van Rij, N. J. W., and Harder, W.,** Microbodies in methanol-assimilating yeasts, *Arch. Microbiol.*, 102, 41, 1975.
54. **Fukui, S., Kawamoto, S., Yasuhara, S., and Tanaka, A.,** Microbody of methanol-grown yeasts. Localization of catalase and flavin-dependent alcohol oxidase in the isolated microbody, *Eur. J. Biochem.*, 59, 561, 1975.
55. **Veenhuis, M., Zwart, K., and Harder, W.,** Degradation of peroxisomes after transfer of methanol-grown *Hansenula polymorpha* into glucose-containing media, *FEMS Microbiol. Lett.*, 3, 21, 1978.
56. **Tsubouchi, J., Tonomura, K., and Tanaka, K.,** Ultrastructure of microbodies of methanol-assimilating yeasts, *J. Gen. Appl. Microbiol.*, 22, 131, 1976.
57. **Osumi, M. and Torigata, S.,** Surface structure of microbodies isolated from yeast cells, in *Scanning Electron Microscopy*, Vol. 2, Proc. Workshop Other Biol. Appl. SEM/STEM, IIT Res. Inst. Chicago, 1977, 617.
58. **Osumi, M., Sato, M., Sakai, T., and Suzuki, M.,** Fine structure of crystalloid in the yeast *Kloeckera* microbodies, *J. Electron Microsc.*, 28, 295, 1979.
59. **Hosoi, J. and Osumi, M.,** Alternate arrangement of large and small particles in crystalloid of *Kloeckera* yeast microbody of halite structure type, *J. Electron Microsc.*, 30, 158, 1981.
60. **Hosoi, J. and Osumi, M.,** Confirmation of composite crystal model of the crystalloid of *Kloeckera* microbody by optical transforms, *J. Electron Microsc.*, 30, 321, 1981.
61. **Osumi, M., Nagano, M., Yamada, N., Hosoi, J., and Yanagida, M.,** Three-dimensional structure of the crystalloid in the microbody of *Kloeckera* sp.: composite crystal model, *J. Bacteriol.*, 151, 376, 1982.
62. **Veenhuis, M., Harder, W., van Dijken, J. P., and Mayer, F.,** Substructure of crystalline peroxisomes in methanol-grown *Hansenula polymorpha:* evidence for an *in vivo* crystal of alcohol oxidase, *Mol. Cell. Biol.*, 1, 949, 1981.
63. **Osumi, M., Nagano, M., and Yanagida, M.,** Structure of paracrystalline arrays in the cell membrane of yeasts, *J. Electron Microsc.*, 28, 301, 1979.
64. **Tani, Y., Miya, T., Nishikawa, H., and Ogata, K.,** The microbial metabolism of methanol. I. Formation and crystallization of methanol-oxidizing enzyme in a methanol-utilizing yeast, *Kloeckera* sp. No. 2201, *Agric. Biol. Chem.*, 36, 68, 1972.
65. **Tani, Y., Miya, T., and Ogata, K.,** The microbial metabolism of methanol. II. Properties of crystalline alcohol oxidase from *Kloeckera* sp. No. 2201, *Agric. Biol. Chem.*, 36, 76, 1972.
66. **Kraus, A. and Simon, H.,** Stereochemistry and mechanism of the methanol oxidase from *Candida boidinii, Hoppe-Seyler's Z. Physiol. Chem.*, 356, 1477, 1975.
67. **Kato, N., Omori, Y., Tani, Y., and Ogata, K.,** Alcohol oxidase of *Kloeckera* sp. and *Hansenula polymorpha*. Catalytic properties and subunit structures, *Eur. J. Biochem.*, 64, 341, 1976.
68. **Fujii, T. and Tonomura, K.,** Oxidation of methanol, formaldehyde and formate by a *Candida* species, *Agric. Biol. Chem.*, 36, 2297, 1972.
69. **Yamada, H., Shin, K.-C., Kato, N., Shimizu, S., and Tani, Y.,** Purification and characterization of alcohol oxidase from *Candida* 25-A, *Agric. Biol. Chem.*, 43, 877, 1979.
70. **Couderc, R. and Baratti, J.,** Oxidation of methanol by the yeast, *Pichia pastoris*. Purification and properties of the alcohol oxidase, *Agric. Biol. Chem.*, 44, 2279, 1980.
71. **Geissler, J. and Hemmerich, P.,** Yeast methanol oxidase: unusual type of flavoprotein, *FEBS Lett.*, 126, 152, 1981.
72. **Mincey, T., Tayrien, G., Mildvan, A. S., and Abeles, R. H.,** Presence of a flavin semiquinone in methanol oxidase, *Proc. Natl. Acad. Sci. U.S.A.*, 77, 7099, 1980.
73. **Fujii, T. and Tonomura, K.,** Oxygen-dependent oxidation of formaldehyde by cell-free extract of a methanol-utilizing yeast, *Candida* sp., *Agric. Biol. Chem.*, 38, 1413, 1974.
74. **van Dijken, J. P. and Harder, W.,** Growth yields of microorganisms on methanol and methane. A theoretical study, *Biotechnol. Bioeng.*, 17, 15, 1975.
75. **Sahm, H.,** Oxidation of formaldehyde by alcohol oxidase of *Candida boidinii, Arch. Microbiol.*, 105, 179, 1975.
76. **Kato, N., Tani, Y., and Ogata, K.,** Enzyme system for methanol oxidation in yeasts, *Agric. Biol. Chem.*, 38, 675, 1974.
77. **Shimizu, S., Ishida, M., Tani, Y., and Ogata, K.,** Flavin changes of *Kloeckera* sp. No. 2201 during adaptation to methanol, *Agric. Biol. Chem.*, 41, 423, 1977.

78. **Shimizu, S., Ishida, M., Kato, N., Tani, Y., and Ogata, K.,** Derepression of FAD pyrophosphorylase and flavin changes during growth of *Kloeckera* sp. No. 2201 on methanol, *Agric. Biol. Chem.*, 41, 2215, 1977.
79. **Eggeling, L., Sahm, H., and Wagner, F.,** Induction of FMN adenylyltransferase in the methanol utilizing yeast *Candida boidinii, FEMS Microbiol. Lett.*, 1, 205, 1977.
80. **Roggenkamp, R., Sahm, H., and Wagner, F.,** Microbial assimilation of methanol. Induction and function of catalase in *Candida boidinii, FEBS Lett.*, 41, 283, 1974.
81. **Yasuhara, S., Kawamoto, S., Tanaka, A., Osumi, M., and Fukui, S.,** Induction of catalase activity in a methanol-utilizing yeast, *Kloeckera* sp. No. 2201, *Agric. Biol. Chem.*, 40, 1771, 1976.
82. **Fujii, T. and Tonomura, K.,** Purification and properties of catalase from a methanol-utilizing yeast, *Candida* sp. N-16, *Agric. Biol. Chem.*, 39, 1891, 1975.
83. **Fujii, T., and Tonomura, K.,** Oxidation of methanol and formaldehyde by a system containing alcohol oxidase and catalase purified from *Candida* sp. N-16, *Agric. Biol. Chem.*, 39, 2325, 1975.
84. **van Dijken, Otto, R., and Harder, W.,** Oxidation of methanol, formaldehyde and formate by catalase purified from methanol-grown *Hansenula polymorpha, Arch. Microbiol.*, 106, 221, 1975.
85. **Kato, N., Tamaoki, T., Tani, Y., and Ogata, K.,** Purification and characterization of formaldehyde dehydrogenase in a methanol-utilizing yeast, *Kloeckera* sp. No. 2201, *Agric. Biol. Chem.*, 36, 2411, 1972.
86. **Sahm, H. and Wagner, F.,** Microbial assimilation of methanol. Properties of formaldehyde dehydrogenase and formate dehydrogenase from *Candida boidinii, Arch. Mikrobiol.*, 90, 263, 1973.
87. **Schütte, H., Flossdorf, J., Sahm, H., and Kula, M.-R.,** Purification and properties of formaldehyde dehydrogenase and formate dehydrogenase from *Candida boidinii, Eur. J. Biochem.*, 62, 151, 1976.
88. **van Dijken, J. P., Oostra-Demkes, G. J., Otto, R., and Harder, W.,** *S*-Formylglutathione: the substrate for formate dehydrogenase in methanol-utilizing yeasts, *Arch. Microbiol.*, 111, 77, 1976.
89. **Kato, N., Sakazawa, C., Nishizawa, T., Tani, Y., and Yamada, H.,** Purification and characterization of *S*-formylglutathione hydrolase from a methanol-utilizing yeast, *Kloeckera* sp. No. 2201, *Biochim. Biophys. Acta*, 611, 323, 1980.
90. **Neben, I., Sahm, H., and Kula, M.-R.,** Studies on an enzyme, *S*-formylglutathione hydrolase, of the dissimilatory pathway of methanol in *Candida boidinii, Biochim. Biophys. Acta*, 614, 81, 1980.
91. **Kato, N., Kano, M., Tani, Y., and Ogata, K.,** Purification and characterization of formate dehydrogenase in a methanol-utilizing yeast, *Kloeckera* sp. No. 2201, *Agric. Biol. Chem.*, 38, 111, 1974.
92. **Hou, C. T., Patel, R. N., Laskin, A. I., and Barnabe, N.,** NAD-linked formate dehydrogenase from methanol-grown *Pichia pastoris* NRRL-Y-7556, *Arch. Biochem. Biophys.*, 216, 296, 1982.
93. **Kato, N., Sahm, H., Schütte, H., and Wagner, F.** Purification and properties of glucose-6-phosphate dehydrogenase and 6-phosphogluconate dehydrogenase from a methanol-utilizing yeast, *Candida boidinii, Biochim. Biophys. Acta*, 566, 1, 1979.
94. **Bykovskaya, S. V. and Trotsenko, Yu. A.,** Metabolism of the methylotrophic yeast *Candida methylica, Microbiology*, 49, 543, 1981.
95. **Mehta, R. J.,** Demonstration of methanol dehydrogenase in methanol assimilating yeasts, *Experientia*, 31, 407, 1975.
96. **Mehta, R. J.,** Pyridine nucleotide-linked oxidation of methanol in methanol-assimilating yeasts, *J. Bacteriol.*, 124, 1165, 1975.
97. **Hou, C. T., Patel, R. N., Laskin, A. I., Barnabe, N., and Marczak, I.,** Microbial oxidation of gaseous hydrocarbons. Production of methyl ketones from their corresponding secondary alcohols by methane- and methanol-grown microbes, *Appl. Environ. Microbiol.*, 38, 135, 1979.
98. **Patel, R. N., Hou, C. T., Laskin, A. I., Derelanko, P., and Felix, A.,** Oxidation of secondary alcohols to methyl ketones by yeasts, *Appl. Environ. Microbiol.*, 38, 219, 1979.
99. **Patel, R. N., Hou, C. T., Laskin, A. I., Derelanko, P., and Felix, A.,** Microbial production of methyl ketones. Purification and properties of a secondary alcohol dehydrogenase from yeast, *Eur. J. Biochem.*, 101, 401, 1979.
100. **Hou, C. T., Patel, R. N., Laskin, A. I., Barnabe, N., and Marczak, I.,** Identification and purification of a nicotinamide adenine dinucleotide-dependent secondary alcohol dehydrogenase from C_1-utilizing microbes, *FEBS Lett.*, 101, 179, 1979.
101. **Hou, C. T., Patel, R. N., Laskin, A. I., Barnabe, N., and Marczak, I.,** Substrate specificity and stereospecificity of nicotinamide adenine dinucleotide-linked alcohol dehydrogenases from methanol-grown yeasts, *Appl. Environ, Microbiol.*, 41, 829, 1981.
102. **Hou, C. T., Patel, R. N., and Barnabe, N.,** Identification of an NADH-linked formaldehyde-reducing enzyme from methanol-grown *Pichia pastoris* NRRL Y-7556, *FEMS Microbiol. Lett.*,, 15, 159, 1982.
103. **Middelhoven, W. J., Berends, J., Repelius, C., and van Aert, A. J. M.,** Excessive production of ammonium from nitrate by some methanol-assimilating yeast strains, *Eur. J. Appl. Microbiol.*, 2, 169, 1976.
104. **van Dijken, J. P., Veenhuis, M., Vermeulen, C. A., and Harder, W.,** Cytochemical localization of catalase activity in methanol-grown, *Hansenula polymorpha, Arch. Microbiol.*, 105, 261, 1975.

105. **Voříšek, J. and Volfová, O.**, Catalase activity in methanol-oxidizing *Candida boidinii* 11Bh and its cytochemical localization, *FEBS Lett.*, 52, 246, 1975.
106. **Osumi, M. and Sato, M.**, Further studies on the localization of catalase in yeast microbodies by ultrathin frozen sectioning, *J. Electron Microsc.*, 27, 127, 1978.
107. **Tanaka, A., Yasuhara, S., Osumi, M., and Fukui, S.**, Immobilization of yeast microbodies by inclusion with photo-crosslinkable resins, *Eur. J. Biochem.*, 80, 193, 1977.
108. **Tanaka, A., Yasuhara, S., Kawamoto, S., Fukui, S., and Osumi, M.**, Development of microbodies in the yeast *Kloeckera* growing on methanol., *J. Bacteriol.*, 126, 919, 1976.
109. **Veenhuis, M., van Dijken, J. P., Pilon, S. A. F., and Harder, W.**, Development of crystalline peroxisomes in methanol-grown cells of the yeast *Hansenula polymorpha* and its relation to environmental conditions, *Arch. Microbiol.*, 117, 153, 1978.
110. **Bormann, C. and Sahm, H.**, Degradation of microbodies in relation to activities of alcohol oxidase and catalase in *Candida boidinii*, *Arch. Microbiol.*, 117, 67, 1978.
111. **Eggeling, L. and Sahm, H.**, Derepression and partial insensitivity to carbon catabolite repression of the methanol dissimilating enzymes in *Hansenula polymorpha*, *Eur. J. Appl. Microbiol. Biotechnol.*, 5, 197, 1978.
112. **Egli, T., van Dijken, J. P., Veenhuis, M., Harder, W., and Fiechter, A.**, Methanol metabolism in yeasts: regulation of the synthesis of catabolic enzymes, *Arch. Microbiol.*, 124, 115, 1980.
113. **Eggeling, L. and Sahm, H.**, Regulation of alcohol oxidase synthesis in *Hansenula polymorpha:* oversynthesis during growth on mixed substrates and induction by methanol, *Arch. Microbiol,*. 127, 119, 1980.
114. **Egli, T., Haltmeier, T., and Fiechter, A.**, Regulation of the synthesis of methanol oxidizing enzymes in *Kloeckera* sp. 2201 and *Hansenula polymorpha*, a comparison, *Arch. Microbiol.*, 131, 174, 1982.
115. **Veenhuis, M., Keizer, I., and Harder, W.**, Characterization of peroxisomes in glucose-grown *Hansenula polymorpha* and their development after the transfer of cells into methanol-containing media, *Arch. Microbiol.*, 120, 167, 1979.
116. **Eggeling, L. and Sahm, H.**, Enhanced utilization-rate of methanol during growth on a mixed substrate: a continuous culture study with *Hansenula polymorpha*, *Arch. Microbiol.*, 130, 362, 1981.
117. **Egli, T.**, Regulation of protein synthesis in methylotrophic yeasts: repression of methanol dissimilating enzymes by nitrogen limitation, *Arch. Microbiol.*, 131, 95, 1982.
118. **Kato, N., Sahm, H., and Wagner, F.**, Steady-state kinetics of formaldehyde dehydrogenase and formate dehydrogenase from a methanol-utilizing yeast, *Candida boidinii*, *Biochim. Biophys. Acta*, 566, 12, 1979.
119. **Bajpai, R. K., Reuss, M., and Held, W.**, Regulation phenomena in methanol consuming yeasts: an experiment with model discrimination, *Biotechnol. Bioeng.*, 23, 499, 1981.
120. **Fujii, T. and Tonomura, K.**, Hexose-phosphate as an intermediate in the assimilation of methanol by *Candida* sp., *Agric. Biol. Chem.*, 37, 447, 1973.
121. **Fujii, T., Asada, Y., and Tonomura, L.**, Assimilative pathway of methanol in *Candida* sp. Incorporation of ^{14}C-methanol, ^{14}C-formaldehyde, ^{14}C-formate and ^{14}-C-bicarbonate into cell constituents, *Agric. Biol. Chem.*, 38, 1121, 1974.
122. **Diel, F., Held, W., Schlanderer, G., and Dellweg, H.**, Comparative investigations on the metabolism of formaldehyde in the presence of ribose-5-phosphate in cell-free extracts of yeasts grown on methanol, *FEBS Lett.*, 38, 274, 1974.
123. **Sahm, H. and Wagner, F.**, Microbial assimilation of methanol. Incorporation of formaldehyde into fructose- and glucose-phosphates by cell-free extract of *Candida boidinii*, *Arch. Microbiol.*, 97, 163, 1974.
124. **Fujii, T. and Tonomura, K.**, Incorporation of ^{14}C-formaldehyde into hexose phosphate by cell-free extract of a methanol-utilizing yeast, *Candida* sp., *Agric. Biol. Chem.*, 38, 1763, 1974.
125. **Kemp, M. B.**, Hexose phosphate synthase from *Methylococcus capsulatus* makes D-*arabino*-3-hexulose phosphate, *Biochem. J.*, 139, 129, 1974.
126. **Kato, N., Ohashi, H., Hori, T., Tani, Y., and Ogata, K.**, Properties of 3-hexulose phosphate synthase and phospho-3-hexuloisomerase of a methanol-utilizing bacterium, 77a, *Agric. Biol. Chem.*, 41, 1133, 1977.
127. **van Dijken, J. P., Harder, W., Beardsmore, A. J., and Quayle, J. R.**, Dihydroxyacetone: an intermediate in the assimilation of methanol by yeasts?, *FEMS Microbiol. Lett.*, 4, 97, 1978.
128. **Babel, W. and Loffhagen, N.**, Assimilation of methanol by yeasts, a new approach, *Z. Allg. Mikrobiol.*, 19, 299, 1979.
129. **Kato, N., Nishizawa, T., Sakazawa, C., Tani, Y., and Yamada, H.**, Xylulose 5-phosphate-dependent fixation of formaldehyde in a methanol-utilizing yeast, *Kloeckera* sp. No. 2201, *Agric. Biol. Chem.*, 43, 2013, 1979.
130. **Lindley, N. D., Waites, M. J., and Quayle, J. R.**, Simultaneous assay of dihydroxyacetone synthase and transketolase in a methylotrophic yeast grown in continuous culture. A cautionary note, *J. Gen. Microbiol.*, 126, 253, 1981.

131. **Lindley, N. D., Waites, M. J., and Quayle, J. R.,** A modified pulse-labelling technique for the detection of early intermediates in microbial metabolism: detection of [^{14}C]dihydroxyacetone during assimilation of [^{14}C]-methanol by *Hansenula polymorpha, FEMS Microbiol. Lett.*, 8, 13, 1980.
132. **Waites, M. J., Lindley, N. D., and Quayle, J. R.,** Determination of the labelling pattern of dihydroxyacetone and hexose phosphate following a brief incubation of methanol-grown *Hansenula polymorpha* with [^{14}C]methanol, *J.Gen. Microbiol.*, 122, 193, 1981.
133. **Waites M. J. and Quayle, J. R.,** Dihydroxyacetone: a product of xylulose 5-phosphate-dependent fixation of formaldehyde by methanol-grown *Candida boidinii, J. Gen. Microbiol.*, 118, 321, 1980.
134. **Waites, M. J. and Quayle, J. R.,** The interrelationship between transketolase and dihydroxyacetone synthase activities in the methylotrophic yeast *Candida boidinii, J. Gen. Microbiol.*, 124, 309, 1981.
135. **O'Connor, M. L. and Quayle, J. R.,** Pentose phosphate-dependent fixation of formaldehyde by methanol-grown *Hensenula polymorpha* and *Candida boidinii, J. Gen. Microbiol.*, 120, 219, 1980.
136. **Bystrykh, L. V., Sokolov, A. P., and Trotsenko, Y. A.,** Purification and properties of dihydroxyacetone synthase from the methylotrophic yeast *Candida boidinii, FEBS Lett.*, 132, 324, 1981.
137. **Kato, N., Higuchi, T., Sakazawa, C., Nishizawa, T., Tani, Y., and Yamada, H.,** Purification and properties of a transketolase responsible for formaldehyde fixation in a methanol-utilizing yeast, *Candida boidinii (Kloeckera* sp.) No. 2201, *Biochim. Biophys. Acta*, 715, 143, 1982.
138. **Hofmann, K. H. and Babel, W.,** Dihydroxyacetone kinase of methanol-assimilating yeasts. I. Regulation of dihydroxyacetone kinase from *Candida methylica in situ, Z. Allg. Mikrobiol.*, 20, 389, 1980.
139. **Hofmann, K. H. and Babel, W.,** Dihydroxyacetone kinase of methanol-assimilating yeasts. II. Partial purification and some properties of dihydroxyacetone kinase from *Candida methylica, Z. Allg. Mikrobiol.*, 21, 219, 1981.
140. **Babel, W. and Hofmann, K. H.,** The relation between the assimilation of methanol and glycerol in yeasts, *Arch. Microbiol.*, 132, 179, 1982.
141. **Babel, W. and Hofmann, K. H.,** The conversion of triosephosphate *via* methylotrophic yeasts?, *FEMS Microbiol. Lett.*, 10, 133, 1981.
142. **Schlanderer, G., Reimann, J., Held, W., and Dellweg, H.,** The isolation of intact mitochondria and studies of oxidative phosphorylation in methanol grown yeasts, Presented at Int. Symp. Microbial Growth on C_1-Compounds, Puschino, September 12 to 16, 1977, 84.
143. **Cooney, C. L. and Levine, D. W.,** SCP production from methanol by yeast, in *Single Cell Protein*, II, Tannenbaum, S. R. and Wang, D. I. C., Eds., MIT Press, Cambridge, 1975, 402.
144. **Kuraishi, M., Matsuda, N., Terao, I., Kamibayashi, A., Tonomura, K., and Fujii, T.,** A study on the internal structure design of air-lift fermentors in the production of methanol single-cell protein, in *Microbial Growth on C_1-Compounds*, The Organizing Committee, Ed., Maruzen, Tokyo, 1975, 231.
145. **Kuraishi, M., Ohkouchi, H., Matsuda, N., and Terao, I.,** A study on the performance of air-lift fermentors in the production of methanol single cell protein, Presented at Int. Symp. Microbial Growth on C_1-Compounds, Puschino, September 12 to 16, 1977, 180.
146. **Denenu, E. O. and Demain, A. L.,** Enzymatic basis for overproduction of tryptophan and its metabolites in *Hansenula polymorpha* mutants, *Appl. Environ. Microbiol.*, 42, 497, 1981.
147. **Tani, Y., Cho, S.-G., Sakai, Y., and Yamada, H.,** unpublished data, 1983.
148. **Shimizu, S., Ishida, M., Tani, Y., and Ogata, K.,** Production of flavin adenine dinucleotide by methanol-utilizing yeasts, *J. Ferment. Technol.*, 55, 630, 1977.
149. **Tani, Y., Mitani, Y., and Yamada, H.,** Utilization of C_1-compounds: phosphorylation of adenylate by oxidative phosphorylation in *Candida boidinii (Kloeckera* sp.) No. 2201, *Agric. Biol. Chem.*,, 46, 1097, 1982.
150. **Baratti, J., Couderc, R., Cooney, C. L., and Wang, D. I. C.,** Preparation and properties of immobilized methanol oxidase, *Biotechnol. Bioeng.*, 20, 333, 1978.
151. **Couderc, R. and Baratti, J.,** Immobilized yeast cells with methanol oxidase: preparation and enzymatic properties, *Biotechnol. Bioeng.*, 22, 1155, 1980.
152. **Tanaka, A., Yasuhara, S., Gellf, G., Osumi, M., and Fukui, S.,** Immobilization of yeast microbodies and the properties of immobilized microbody enzymes, *Eur. J. Appl. Microbiol. Biotechnol.*, 5, 17, 1978.
153. **Yamada, H., Kishimoto, N., and Kumagai, H.,** Metabolism of N-substituted amines by yeasts, *J. Ferment. Technol.*, 54, 726, 1976.
154. **van Dijken, J. P. and Bos, P.,** Utilization of amines by yeasts, *Arch. Microbiol.*, 128, 320, 1981.
155. **Zwart, K., Veenhuis, M., van Dijken, J. P., and Harder, W.,** Development of amine oxidase-containing peroxisomes in yeasts during growth on glucose in the presence of methylamine as the sole source of nitrogen, *Arch. Microbiol.*, 126, 117, 1980.
156. **Veenhuis, M., Zwart, K. B., and Harder, W.,** Biogenesis and turnover of peroxisomes involved in the concurrent oxidation of methanol and methylamine in *Hansenula polymorpha, Arch. Microbiol.*, 129, 35, 1981.
157. **Tani, Y., Mori, N., Ogata, K., and Yamada, H.,** Production and purification of choline oxidase from *Cylindrocarpon didymum* M-1, *Agric. Biol. Chem.*,, 43, 815, 1979.

158. **Tani, Y., Mori, N., and Ogata, K.,** A new fungal enzyme, choline oxidase, *Agric. Biol. Chem.*, 41, 1101, 1977.
159. **Yamada, H., Mori, N., and Tani, Y.,** Properties of choline oxidase of *Cylindrocarpon didymum* M-1, *Agric. Biol. Chem.*, 43, 2173, 1979.
160. **Mori, N., Kawakami, B., Hyakutome, K., Tani, Y., and Yamada, H.,** Characterization of betaine aldehyde dehydrogenase from *Cylindrocarpon didymum* M-1, *Agric. Biol. Chem.*, 44, 3015, 1980.
161. **Mori, N., Kawakami, B., Tani, Y., and Yamada, H.,** Purification and properties of dimethylglycine oxidase from *Cylindrocarpon didymum* M-1, *Agric. Biol. Chem.*, 44, 1383, 1980.
162. **Mori, N., Sano, M., Tani, Y., and Yamada, H.,** Purification and properties of sarcosine oxidase from *Cylindrocarpon didymum* M-1, *Agric. Biol. Chem.*, 44, 1391, 1980.
163. **Tye, R. J. and Willetts, A. J.,** Fungal growth on methanol, *J. Gen. Microbiol.*, 77, i, 1973.
164. **Willets, A.,** Fungal growth on C_1 compounds: quantitative aspects of growth of a methanol-utilizing strain of *Trichoderma lignorum* in batch culture, *Appl. Environ. Microbiol.*, 33, 758, 1977.
165. **Sakaguchi, K., Kurane, R., and Murata, M.,** Assimilation of formaldehyde and other C_1-compounds by *Gliocladium deliquescens* and *Paecilomyces varioti*, *Agric. Biol. Chem.*, 39, 1695, 1975.
166. **Wolf, H. J. and Hanson, R. S.,** Isolation and characterization of methane-utilizing yeasts, *J. Gen. Microbiol.*, 114, 187, 1979.
167. **Wolf, H. J. and Hanson, R. S.,** Identification of methane-utilizing yeasts, *FEMS Microbiol. Lett.*, 7, 177, 1980.
168. **Wolf, H. J., Christiansen, M., and Hanson, R. S.,** Ultrastructure of methanotrophic yeasts, *J. Bacteriol.*, 141, 1340, 1980.
169. **Bringer, S., Sprey, B., and Sahm, H.,** Purification and properties of alcohol oxidase from *Poria contigua*, *Eur. J. Biochem.*, 101, 563, 1979.
170. **Takahashi, S., Itoh, M., Tsubaki, K., and Kaneko, Y.,** Taxonomical identification of phenol- and *o*-cresol-assimilating fungus *Aureobasidium pullulans* and its growth characteristics in phenol medium with methanol or formaldehyde, *Agric. Biol. Chem.*, 45, 1809, 1981.
171. **Kato, N., Miyawaki, N., and Sakazawa, C.,** Oxidation of formaldehyde by resistant yeasts, *Debaryomyces vanriji* and *Trichosporon penicillatum*, *Agric. Biol. Chem.*, 46, 655, 1982.

Chapter 3

GENETICS OF METHYLOTROPHS

B. W. Holloway

TABLE OF CONTENTS

I. INTRODUCTION

Forty years of *Escherichia coli* K12 genetics has established bacterial genetics as an essential component of microbiological studies. The sophisticated and extensive knowledge of this organism has been essential for the establishment and success of molecular biology and recombinant DNA technology. An important aspect of this has been to demonstrate to microbiologists and biochemists the value of genetic analysis as a means to the solution of a wide variety of biological problems.

The use of specific mutants to determine pathways of intermediary metabolism, and the analysis of microbial development[1] are typical examples. As better procedures for mutagenesis, exchange of genetic material and the use of transposable genetic units have developed, the advantages of genetic analysis have become available to an increasing number of bacterial genera.[2,3] As a result there is a wider understanding of microbial genetics, new genetic phenomena have been discovered, and the biology of these organisms is now better known.

Methylotrophic bacteria have been recognized and studied for nearly 80 years but it is only over the last 10 years that efforts have been made to develop means of genetic analysis for them. The aim of this review is to record progress to date and to indicate areas where future successes might lie.

II. THE AIMS OF GENETIC ANALYSIS IN BACTERIA

Although extensive genetic analysis has been limited to less than 10 species, the genetic structure of various bacteria shows many common features. The main genomic structure is a circular double stranded DNA chromosome. The size of this chromosome has been measured by sedimentation analysis, contour length, and DNA denaturation kinetics, and there is a tendency for the chromosome of various bacteria to be clustered into size groups around 750, 2100, and 4000 kb.[4,5]

The size of DNA fragments is measured either in molecular units (daltons) or as kilobases of double stranded DNA. The advantage of using the kilobase system is that 1 kb is approximately the amount of DNA needed to code for one protein, so that the functional potential of a segment of DNA can be more readily assessed.

In addition, there are extrachromosomal, self-replicating genetic units, plasmids, which may range in size from 2.3 kb to over 300 kb. The knowledge of characteristics identified as being encoded by plasmids is being continually extended and currently includes antibiotic resistance, metal resistance, substrate utilization, virulence, nitrogen fixation, host chromosome mobilization, and plant tumor formation.[6] To date, no plasmid has been identified which codes for a function essential to the growth of the bacterium which carries it. The stability of plasmids carried by bacteria varies greatly, depending upon the growth environment and genetic features of the plasmid and the host. However, it is not uncommon for all the plasmid DNA to be lost from a bacterium. In such a case, while the selective growth ability of the original bacterium may be reduced for given environments, the organism which has lost the plasmid can still survive and grow in less rigorous environments. This feature may have significance for the genomic location of genes for the utilization of C_1 compounds in obligate methylotrophs.

Bacteriophages are also associated with most bacteria, and these bacterial viruses have specific host requirements for their replication. Bacteriophages may have either RNA or DNA as their prime genetic substance and if DNA, it may be capable of physical integration into the bacterial chromosome, a state known as lysogeny. Some bacteriophages can transfer fragments of bacterial chromosome during infection, a process known as transduction. Bacteriophages are an important genetic component of bacteria, and transduction is a very precise means of fine genetic mapping.

The genetic analysis of bacteria may be accomplished by either physical or biological techniques; ideally the two should be combined. The physical isolation of whole bacterial chromosomes is difficult and studies using the appropriate techniques to do this have been important for understanding the mechanisms of bacterial chromosome replication. The physical analysis of plasmids, in terms of isolation of DNA and its characterization by restriction endonuclease and hybridization techniques, is well established and applicable to plasmids in general. With some bacterial species, the established techniques for isolation of plasmid DNA may need adaptation, common problems being the ease with which the bacterial cells lyse and the separation of plasmid DNA from other cellular components.

The first step in the classical biological techniques of genetic analysis involves the isolation of a range of mutants affecting different functions. The success of such mutant isolation generally requires an effective selection procedure. The choice of the most appropriate mutants is usually based on a knowledge of the growth characteristics of the organism under study. The frequency of such mutations can be increased by the use of mutagens which increase the spontaneous mutation rate. Mutants which have been extensively used in various bacteria include auxotrophs, antibiotic resistance, resistance to other toxic agents, metabolic deficiencies (for example, lacking the ability to use a particular substrate or ferment a carbohydrate), suppressor mutants, and cell wall mutants. A class of mutant which has been particularly useful, and could have useful applications for methylotrophs, is the conditional lethal mutant. These mutants display a mutant phenotype under certain specified environmental conditions and a wild-type phenotype under others. The most common type of conditional lethal mutant is the temperature-sensitive mutant; mutant at one growth temperature (say 40°C) and wild-type at another temperature (say 28°C). Such mutants have already played a role in the genetic manipulation of an obligate methylotroph.[7]

Once mutants have been obtained, the next step in the genetic analysis is to locate the site of the gene which has been mutated. It may be situated either on the chromosome or located on a plasmid. Distinguishing between these two locations may be difficult and the success of doing so will depend upon the level of genetic knowledge of the organism and the nature of the plasmid. An instability of the mutant phenotype higher than that commonly found with chromosome mutations (the common spontaneous reversion rate of chromosomal genes is about 10^{-8}) and an increased frequency loss of the phenotype following treatment with mitomycin C may be evidence for a plasmid encoded function. Further evidence for the location of a particular gene on a plasmid may be provided by conjugation experiments in which the characteristic is transferred from a donor strain to a recipient strain at frequencies characteristic of plasmid transfer (10^{-1} to 10^{-4}/donor parent cell). However, plasmids may be transfer- or mobilization-deficient and in such circumstances the problems of differentiating the location of a gene between chromosome and plasmid become more difficult.

The importance of establishing a comprehensive chromosome map for a bacterium cannot be overemphasized. It provides the basic data necessary for the genetic manipulation of an organism by recombinant DNA techniques or the construction of strains with desired phenotypic properties by more conventional genetic means. It aids the differentiation of mutants with similar phenotypic properties but which are genetically different. It may provide insights into the mechanism of regulation of a metabolic pathway as demonstrated in the case with operons with a number of bacteria.

Chromosomal mapping of bacteria is best achieved by conjugation, this invariably being plasmid-mediated. Alternative effective procedures are transduction, in which a bacteriophage transfers short segments of bacterial chromosome and transformation, in which DNA fragments from a donor bacterium enter a recipient cell and become integrated into the chromosome. There is now a variety of plasmids which promotes chromosome mobilization for a variety of bacteria.[3] Two types of plasmids deserve special mention in this respect. The first type are those derived from certain IncP-1 plasmids which have a wide bacterial host range and carry specific DNA sequences which promote chromosome transfer in many different bacteria. The best known example is R68.45[8,9] and this plasmid will be discussed in detail below.

The second type of plasmids are those which carry a transposon which can be translocated from the plasmid to the bacterial chromosome. This provides a region of genetic homology between the plasmid and the chromosome which enables the plasmid to mobilize the bacterial chromosome at conjugation.[10,11] Both these types of plasmids have considerable potential for the development of conjugation systems in methylotrophs.

A highly effective combination of the physical and biological systems is found in the use of recombinant DNA techniques. Fragments of bacterial chromosome, produced by the action of restriction endonucleases, can be incorporated into selected plasmid vectors which have themselves been treated by the same enzyme. Ideally, the plasmid vector should have only one site for each enzyme, so that the plasmid vector DNA is linearized by the enzyme, the bacterial fragment can then be inserted at the unique site, after which the plasmid DNA now containing the bacterial DNA fragment can be recircularized by ligase treatment. Difficulties arise in selecting a particular region or gene of the bacterial chromosome for cloning in this way and a suitable selective technique is always required. Gene banks have now been prepared for a number of bacteria.[12] The DNA of various bacteria are usually cloned in *E. coli* because the widest range of cloning vectors is available for this bacterium. However, difficulties may arise in that not all genes from other bacteria are expressed in *E. coli* and other plasmid vectors and bacterial hosts are becoming available for cloning purposes. There are as yet no cloning vectors specifically developed for methylotrophs.

III. GENETICS OF METHYLOTROPHS

The genetics of *E. coli* was established using essentially one strain, K12. A similar limited approach of concentrating genetic studies on one or a few strains has also occurred with *Salmonella typhimurium*, *Pseudomonas aeruginosa*, and *Bacillus subtilis*. With methylotrophs, however, the genetic effort has been diluted over a number of organisms, both obligate and facultative methylotrophs, and there is little consensus of results which can be applied to the group as a whole. The following description of genetic studies in methylotrophs is thus of necessity fragmented and it is to be hoped that a few organisms will commend themselves to workers in this area for intensive genetic study.

A. Mutation

Isolation of mutants in methylotrophs is demonstrably less easy than is the case with the Enterobacteriaceae or the pseudomonads. For example, Williams et al.[13,108] with the methylotrophs *Methylococcus capsulatus* and *Methylomonas albus* found that ultraviolet irradiation (UV), gamma radiation, methyl methane sulfonate (MMS), and ethyl methane sulfonate (EMS) did not induce streptomycin-resistant mutants in either organism, but *N*-methyl-*N'*-nitro-*N*-nitrosoguanidine (NG) did so in *M. albus,* although at lower frequency than in *E. coli.* By contrast in *Methylobacterium organophilum,* O'Connor et al.[14] were able to isolate mutants using both UV and NG. The variants identified included auxotrophs, drug-resistant mutants and mutants unable to grow on methane or methanol but capable of growth on glucose or succinate. Windass el al.[7] successfully used NG with *Methylophilus methylotrophs* for the isolation of temperature-sensitive mutants including four which had lost GOGAT function.

There is insufficient data from the variety of organisms used to draw any general conclusions as to why it is difficult to isolate mutants of methylotrophs. It has been suggested that the lack of error prone "SOS" DNA repair mechanisms may be responsible for their lack of response in this respect to UV radiation.[15,16] The permeability to mutagens may be diminished in methylotrophs and the specific growth conditions needed to grow these organisms may affect the efficiency of the mutagens tested. One chemical mutagen which does not seem to have been used is manganous chloride, shown to be an effective mutagen in both *E. coli* and *Pseudomonas aeruginosa.*[17,18] There is no evidence that induced mutants of methylotrophs are any more stable or unstable than those found in other bacteria. However, permeability properties of methylotrophs in artificial culture could create selective conditions under which some revertants had growth advantages. It would be profitable to carry out systematic studies on the effectiveness of a range of mutagens for a limited range of selected methylotrophs.

A new technique which has proved to be successful for the isolation of mutants in a variety of microorganisms is that of transposon mutagenesis. Transposable elements carrying antibiotic resistance determinants can transpose from the plasmids carrying them to sites on bacterial chromosomes or onto other plasmids carried by the same bacterium. Insertion of the transposon may result in a mutation of the gene into which it has inserted and the presence of the antibiotic resistance determinant of the transposon provides a selective marker for the mutation so caused. The effectiveness of transposon mutagenesis depends upon the transposon used and the degree of selection which is available for the transpositional event. The range of methods and uses have been described in detail.[19] The type of procedure which would be applicable to methylotrophs involves the creation of a ''suicide'' situation in which a plasmid carrying the transposon is transferred to a recipient methylotroph, selection being imposed for the antibiotic resistance coded for by the transposon. The plasmid chosen to carry the transposon cannot replicate in the methylotroph. Thus selection can be made for those events in which the transposon is mobilized from the plasmid into a chromosomal site. Survivors carrying the transposon, as identified by their antibiotic resistance pattern, can then be screened for the chromosomal mutation sought. One such plasmid, pJB4JI, was constructed by Beringer et al.[20] It carries Tn*5* conferring resistance to kanamycin, and is an RP4(IncP-1)-Mu hybrid, the Mu itself being inserted into RP4 and preventing establishment of the RP4 plasmid in various hosts. It has been successfully used by O'Connor[21] to isolate mutants of *Xanthobacter* H4.14.

Sato et al.[22] have developed a host-dependent hybrid plasmid which is suitable as a suicide carrier for transposons. pAS8 Rep-1 is a chimeric plasmid containing a ColE1 replicon and RP4 conjugation genes. The latter enables the plasmid to have a wide host range, but ColE1 plasmids will only replicate in a limited range of species which does not include methylotrophs. This plasmid has been successfully used in phytopathogenic *Pseudomonas*[22] and in *Bordetella pertussis.*[11] Undoubtedly other suicide plasmids can be constructed which will be even better suited to methylotrophs. A search should be made for plasmids which have a limited host range not including methylotrophs and which can be loaded with various transposons. One possibility is R91-5, an IncP-10 plasmid with a host range restricted to *P. aeruginosa.*[23] Identification of transposons which can express antibiotic resistance and show low specificity of insertion will aid effective transposon mutagenesis in these organisms. Both Tn*5* and Tn*501* (the latter coding for mercury resistance) seem to be sufficiently nonspecific in terms of their insertion sites for them to act as general mutagenic elements and at this time would be the transposons of choice.

B. Systems of Gene Exchange

A number of gene exchange systems are being studied in methylotrophs, but to date there is minimal mapping data and the development of effective systems of chromosome exchange should be a high priority for workers in this area.

1. Transduction

As yet there are few reports of transducing bacteriophages for the methylotrophs. Oki et al.[24] were the first to isolate a bacteriophage for any methanol assimilating bacterium. Using *Methanomonas methylovora,* three types of bacteriophages were distinguished. Workers in the German Democratic Republic and the Soviet Union have isolated 23 phage strains active on various methanotrophic bacteria.[25] They all had a long latent period and small burst size, although the plaque morphology is typical of phages isolated from natural sources for other bacteria. There is no obvious reason why attempts to isolate phages for methylotrophs by other workers have not been successful.

There is also no reason why such phages should not be potentially useful for transduction. While for some time it was generally assumed that virulent bacteriophages would be incompetent as transducing phages, this view has not been supported by recent data. E79, a virulent bacteriophage for *P. aeruginosa* can transduce,[26] and it has been shown that T1 and T4, virulent coliphages can transduce chromosomal genes of *E. coli.*[27,28] Every effort should be made to seek bacteriophages for those strains of methylotrophs which are of interest to microbiologists and biochemists. The advantages of having a fine structure mapping procedure which would accrue from transduction would more than justify the effort involved.

2. *Transformation*

Transformation was the earliest form of bacterial genetic analysis and its value continues to be demonstrated in a variety of microorganisms.[2] Like transduction, it is a means of fine structure genetic analysis, although the analysis of linkage is not quite so accurate or direct. Transformation has been demonstrated in *Methylobacterium organophilum*[14] and *Methylococcus capsulatus.*[29] The techniques used have been essentially the same as those developed for other bacterial transformation systems. Lysis of the donor cells was achieved by either sodium dodecyl sulfate or by lysozyme, the chromosomal DNA then being purified by phenol extraction. In both organisms, competence was highest in cells near the end of exponential growth and frequencies of transformation up to 0.5% were obtained, although relatively high concentrations of DNA were required. Such methods are entirely satisfactory for determining linkage of closely linked genes and mapping data has been obtained in *Methylobacterium organophilum* showing linkage of genes concerned with methanol dissimilation.[30,31]

If the benefits of DNA recombinant technology are to become available for the methylotrophs, techniques for plasmid transformation will need to be developed.[6]

3. *Conjugation*

The development of effective systems of transfer of chromosomal DNA should be the most important objective of current genetic research with methylotrophs. In some organisms, such as *E. coli* or *P. aeruginosa,* native plasmids have provided the means of developing chromosomal exchange systems and the search for native plasmids in methylotrophs is important for this reason. Two important goals can be identified in the current search for plasmids in methylotrophic bacteria. The first concerns the role that plasmids play in the location of genetic information needed for the dissimilation of C_1 compounds. The second is the identification of plasmids which can be used for chromosome mobilization studies or act as vectors in recombinant DNA procedures. For the present the second of these goals is being achieved using plasmids isolated in other genera of bacteria.

a. Plasmids of Methylotrophs

Plasmids have been identified in a variety of methylotrophs. The methods so far used in their isolation do not substantially differ from those used for other bacteria. It should be recalled however that the unmodified application of procedures developed for enterobacteria may lead to frustration when used for more diverse microorganisms. Experience has shown that lysis of some methylotrophs to release DNA is more difficult than that found with *E.*

coli. As well, the separation of DNA from cell debris may also be more difficult as was demonstrated in *P. putida* by Palchaudhuri and Chakrabarty.[32]

The first description of plasmids in methylotrophs was made by Warner et al. in 1977.[33,109] They found various plasmids in *Pseudomonas* AM1 with sizes 23 kb, 29 kb, and 32.5 kb. No metabolic function or other phenotypic characteristics have been identified as being encoded by these plasmids. They were unable to identify plasmids in three obligate methylotrophs, *Methylomonas methanica, Methylococcus capsulatus*, and *Methylosinus trichosporium* and one facultative methylotroph, *Pseudomonas extorquens*. The first report of a plasmid in an obligate methylotroph was made by Monteiro et al.[34] *Methylomonas (Methanomonas) methylovora* was shown to carry a 133-kb plasmid and subsequently[35] this plasmid was shown to be conjugative. Stahl and Esser[36] found plasmids in several strains of the obligate methylotroph *Methylomonas clara*. Two related strains were each found to carry a plasmid, in one case 44 kb and in the other 16.6 kb. Restriction endonuclease analysis indicated that the two plasmids were related, suggesting that the smaller one was a deletion product of the larger plasmid. Because neither plasmid was found in another closely related strain of *M. clara* it was concluded that methylotrophy was not encoded by plasmids in this organism.

Plasmids have been identified in *Methylobacterium organophilum* XX and *M. organophilum* CRL26-SB1 and the ability of those strains to grow on methane can be correlated with the presence of 115-kb plasmids.[37] Other obligate methylotrophs have been examined for plasmids. The type I strains, *Methylobacter capsulatus* Y, *Methylomonas methanica* and *M. albus* were shown to have plasmids and four type II strains examined all carried plasmids. Three strains of *Methylosinus trichosporium* each had three plasmids, 147 kb, 150 kb and 160 kb, respectively; *M. sporium* 5 had two plasmids, *Methylocystis parvus* OBBP had two plasmids and *Methylocystis* POC had three. Restriction endonuclease analysis showed that apart from the three strains of *Methylosinus trichosporium* for which the three plasmids seemed to be identical, there was little similarity between the other plasmids. By Southern hybridization,[38] the plasmid DNA from *M. sporium* and *Methylomonas albus* did show some homology.[39]

The diversity of plasmid composition in this range of strains would seem to argue against the genes for C_1 metabolism being located on plasmids. However the situation with facultative methane oxidizers may be different. There is evidence[40] that the ability of such strains to grow on methane is unstable during prolonged laboratory storage. Such instability is often indicative of a plasmid which codes for the phenotype which is lost, as has been shown in *Pseudomonas*.[88] The evidence for plasmid coding of methane oxidation is at present inconclusive.[21]

b. Transfer of Plasmids to Methylotrophs

The second aspect of plasmids and methylotrophs is to determine whether plasmids characterized in other organisms can be transferred to methylotrophs. Several workers have shown this to be possible for the IncP-1 plasmids which are well characterized for their wide bacterial host range. Warner et al.[41] transferred R68.45[8] to *P. extorquens* and *Methylosinus trichosporium* OB3b, and this plasmid was then retransferred from *P. extorquens* to *E. coli, P. aeruginosa*, and *P. extorquens*. Membrane matings show better transfer frequencies than plate matings as it is known that IncP-1 plasmids transfer very inefficiently in liquid mating. In transfers of IncP-1 plasmids to other organisms, more efficient transfer is often obtained by using an *E. coli* donor. This is due to the ability of *E. coli* to methylate DNA and this methylation provides some protection against the native restriction enzymes of the host. At least one methylotroph has been shown to have restriction enzymes. It has been found that *Methylophilus methylotrophus* AS1 has at least two DNA endonucleases, and both nucleases are involved in restriction in vivo. One enzyme has the same recognition sequence as several known restriction enzymes, while the other one has a novel recognition site. Mutants of *M. methylotrophus* AS1 lacking either enzyme are much better recipients for plasmids transferred from other bacteria.[42]

Warner et al.[41] were not able to transfer R68.45 to an obligate *Methylococcus* species (NCIB11083). Difficulties in maintaining IncP-1 plasmids in other bacteria have been reported and in one case the reason is known. Godfrey et al.[43] found that when IncP-1 plasmids were transferred to *P. aeruginosa* PAO they displayed high stability. However, when transferred to a different strain, *P. aeruginosa* PAT, the IncP-1 plasmids R18, R68, and R68.45 showed structural instability, breaking down in successive generations such that the events leading to plasmid instability occurred at a frequency of ca. 10^{-2} per cell per generation. The events leading to instability involved the presence of another plasmid, pVS1 occurring naturally in strain PAT. The demonstration of native plasmids in various methylotrophs means that there could be a similar interaction between such plasmids and IncP-1 plasmids. Jeyaseelan and Guest[44] transferred the IncP-1 plasmid R68.45 to several species of facultative methylotrophs. Transfer to *Pseudomonas* 3A2 occurred at a frequency of 1×10^{-5} per recipient cell when selection was made for kanamycin resistance but selection for carbenicillin resistance was unsatisfactory with this strain and also for *Pseudomonas* AM1, *P. extorquens* and *Methylobacterium organophilum* due to the high frequency of spontaneous carbenicillin-resistant mutants. The frequency of transfer of kanamycin resistance to these three organisms with this plasmid occurred at a very low frequency and no attempts were made to transfer the plasmid from the suspected transconjugants to a suitable recipient to establish that the intact plasmid had survived transfer.

Tatra and Goodwin[45] transferred R68.45 to *Pseudomonas* AM1 from *P. aeruginosa* PAO at a frequency of 10^{-6} per donor cell and Holloway[46] transferred R68.45 to *Methylophilus methylotrophus* AS1. Windass et al.[7] showed that the IncQ plasmid pTB70 could be transferred to *M. methylotrophus* AS1 by mobilization with RP4. They also found that the IncP-1 plasmid pRP301, a deletion derivative of RP4, was transferred readily to *M. methylotrophus*. Care must be taken to use the correct levels of antibiotic resistance in selecting for transfer of plasmids to methylotrophs. There are unfortunately no rules which can help predict the level of expression in a methylotroph of a given antibiotic resistance determinant and the best experimental procedure is to first measure the Minimal Inhibitory Concentration (M.I.C.) of the selected methylotroph to the antibiotic in question.[47] Then, knowing the range of resistances over which the plasmid can confer resistance in bacteria other than methylotrophs, a range of antibiotic levels can then be selected to detect transfer of the plasmid. It should be recalled that not all antibiotic resistances carried by plasmids are necessarily expressed in all bacteria.

Nevertheless, from the above there is no reason to believe that plasmids whose genetic effectiveness in chromosome mobilization has been shown in other genera cannot be transferred for this purpose to methylotrophs. The significance of the IncP-1 plasmids lies in their established ability to mobilize chromosome in a variety of Gram-negative bacteria.[3,48] For the IncQ plasmids, their potential importance lies in their use as cloning vehicles in Gram-negative organisms other than *E. coli*.[7,49]

c. ECM Plasmids

While a number of plasmids have been used to develop chromosome mobilizing systems in a variety of Gram-negative bacteria,[3] the IncP-1 plasmids have attracted particular attention in view of their wide host range, which is essentially most Gram-negative bacteria. It was shown that in some organisms notably *P. aeruginosa* strain PAT,[50] *Acinetobacter calcoaceticus*,[51] *P. glycinea*,[52] *Caulobacter crescentus*,[53] and *Proteus mirabilis*,[54] native IncP-1 plasmids could mobilize bacterial chromosome. However for most other bacteria they were unable to carry out this function and attempts have been made to improve the chromosome mobilizing ability (Cma) of IncP-1 plasmids. Dénarié et al.[55] and Faelen et al.[56] developed techniques for the insertion of Mu into RP4, involving the use of a temperature-sensitive mutant of Mu. Murooka et al.[57] showed that the host range of coliphage Mu could be greatly expanded by using a temperature-sensitive mutant of Mu integrated into RP4 and demonstrated that this system could be used for genetic analysis and intergeneric transfer in a range of Gram-negative bacteria.

Derivatives of the IncP-1 plasmid R68 have been isolated which have acquired Cma for a wide variety of bacteria and of these plasmids, R68.45 is the most widely studied.[3,8,9,58,59] The term Enhanced Chromosome Mobilizing (ECM) plasmids has been applied to this type of plasmid. The Cma of R68.45 depends upon the duplication of a preexisting 2.1 kb segment of DNA already present in R68 to form a tandem structure which is near the kanamycin resistance determinant. This 2.1-kb segment has been shown to have the properties of an insertion element and has been named IS*21*.[60,61] IS*21* does not transpose from R68, but does so with high frequency from R68.45. It has been proposed that the mobilization of chromosome by R68.45 arises from its ability to form a cointegrate when IS*21* transposes to the chromosome. This must occur with high frequency and low site specificity because of the observed frequency of recombinant formation with R68.45 and the multiple origins of transfer in such a wide range of bacteria. IS*21* is entirely a plasmid sequence; to date found only in some IncP-1 plasmids and shown to be absent from the *Pseudomonas aeruginosa*, *P. putida* and *E. coli* chromosomes. Plasmids like R68.45 have been derived from IncP-1 plasmids other than R68 and one has been found in a strain of *P. aeruginosa* isolated in a Japanese hospital. All these plasmids have the same tandem duplication of IS*21*.[60,62]

R68.45 and related ECM plasmids have been shown to effectively mobilize the chromosome of a variety of other Gram-negative bacteria including *Agrobacterium tumefasciens*,[62] *Azospirillum brasilense*,[64] *Erwinia carotovora*,[65] *E. chrysanthemi*,[66] *Escherichia coli*,[67,68] *Klebsiella pneumoniae*,[69] *Rhizobium leguminosarum*,[68] *R. meliloti*,[70] *R. trifolii*,[71] *Rhodopseudomonas capsulata*,[72] *R. sphaeroides*,[73] and *Zymomonas mobilis*.[74]

R68.45 was first shown to mobilize the chromosome of a methylotroph by Holloway using *M. methylotrophus* AS1.[46] Using mutants resistant to streptomycin, nalidixic acid and rifampicin, recombinants were found at a frequency of 10^{-7} per donor parent cell. Tatra and Goodwin[45] using the facultative methylotroph *Pseudomonas* AM1 were able to show that in matings with R68.45 donors there was segregation of auxotrophic markers, genes coding for methanol utilization, and antibiotic resistance markers. Gene transfer frequencies as high as 10^{-4} per donor cell were reported. Linkage of the genes coding for resistance to cycloserine, phosphonomycin and streptomycin to the gene for methanol dehydrogenase was observed.

A range of other ECM plasmids has been isolated.[75] They show variation in stability, ability to mobilize chromosome in various species of *Pseudomonas*[76] and ability to generate prime plasmids (see below). It is likely that R68.45 and other related Cma derivatives of IncP-1 plasmids will be of considerable value in developing conjugative systems for chromosomal analysis in methylotrophs. It would appear that the facultative methylotrophs may be a more promising ground because of the wider range of markers which can be isolated. In this respect, auxotrophs will be of particular value because of their effectiveness as selective markers in conjugation. The availability of auxotrophs means that there are at least 50 markers from the amino acid, purine, and pyrimidine biosynthetic pathways and which can be used to demonstrate location and linkage arrangements of genes for C_1 metabolism and other pathways having special significance for methylotrophs.

What is the prospect for the obligate methylotrophs? The possibilities for obtaining auxotrophs do not seem to be good. In *M. methylotrophus* it is difficult to isolate auxotrophs and hence to obtain the multiply marked strains so essential for comprehensive chromosomal analysis.[77] However ECM plasmids have additional properties which may enable this impasse o be overcome.

d. Transposon Facilitated Recombination

It has been demonstrated that the insertion of some transposons into the bacterial chromosome can provide regions of homology for a plasmid having the same transposon.[19] The advantage of this is that polarized chromosome transfer will occur giving a conjugation linkage analysis system similar to that found with *E. coli* K12 Hfr donors. Krishnapillai et al.[10] translocated Tn*1* into the chromosome of *P. aeruginosa* from the IncP-1 plasmid R18. Twelve different sites of insertion were identified. These insertions served as regions of homology for host chromosome mobilization by either R18 or the IncP-10 plasmid R91-5. Due to the fact that Tn*1* is inserted in opposite orientation in each of these plasmids, chromosome can be mobilized in each direction from each origin site by using one or the other plasmid. Similar systems of chromosome mobilization have been described for *Salmonella typhimurium* with Tn*10*,[89] for *Vibrio cholerae* with Tn*1*,[90] and for *Bordetella pertussis*.[99]

There is a variety of means by which transposons can be inserted into bacterial chromosomes. They usually involve a suicide plasmid which cannot replicate in a given host so that selection for the antibiotic resistance of the transposon must involve its translocation from the plasmid (which is then lost) to a chromosomal site where it is retained as part of the bacterial chromosome. One common way in which this can be done is to have the transposon on a plasmid which has a temperature-sensitive replication or maintenance function. Growth of the organism carrying the plasmid at the nonpermissive temperature combined with selection for the transposon coded antibiotic resistance will allow for isolation of survivors in which the transposon has inserted into the bacterial chromosome. Development of the appropriate plasmid systems in methylotrophs could result in effective gene transfer techniques.

This technique can also be used to construct donors with Hfr-like properties in which the whole plasmid has been inserted into the host bacterial chromosome. Haas et al.[91] isolated a mutant of RP1 temperature sensitive to maintenance such that it could not be maintained in *P. aeruginosa* at 43°C. By growth at this temperature, rare survivors could be isolated which grew at this temperature and retained the whole plasmid which had been integrated into the chromosome. Such survivors could mobilize chromosome from the site of the insertion of the plasmid.

4. Prime Plasmids

In addition to mobilizing the chromosome of various bacteria, R68.45 and other ECM plasmids have the ability to form prime plasmids. These are plasmids which have incorporated a fragment of bacterial DNA into the plasmid genome. The wide host range and antibiotic resistance markers of the parent ECM plasmid are retained so that such primes provide an excellent means of in vivo cloning. They are being used for determining the expression of DNA of one bacterial species in another bacterial species. They provide a first stage enrichment for in vitro cloning of bacterial chromosomal genes, increasing the effectiveness of "shotgun" cloning procedures for such chromosomal genes. Prime plasmids may be formed in one of two ways. Either within a given species, by mating an R68.45 donor to a Rec^- recipient, the latter preventing chromosomal integration of the chromosomal segment of the prime plasmids. Alternatively, another bacterial species can be used as a surrogate recombination deficient recipient, there being little or no integration of such DNA fragments from the donor into the continuity of the bacterial chromosome of the recipient. With either method, the formation of exconjugants carrying prime plasmids is in the range 10^{-7} to 10^{-9} per donor parent cell. Where it has been studied, the molecular structure of prime plasmids derived from R68.45 is that the inserted bacterial DNA is flanked at each of the junctions with the plasmid genome by one copy of IS*21*.

Using these two procedures plasmid primes of R68.45 have been found in *P. aerugi-*

nosa,[78,79] *P. putida*,[80] *E. coli*,[62] *Rhizobium leguminosarum*,[81] *R. meliloti*,[82] *K. pneumoniae*,[83] *Rhodopseudomonas capsulata*,[72] *Acinetobacter calcoaceticus*,[84] and *Methylophilus methlotrophus* AS1.[77]

The primes of *M. methylotrophus* have proved effective for mapping genes on the chromosome of that organism. Attempts to map this organism with R68.45 were ineffective because of the difficulties of isolating auxotrophs and other markers which could be used for the selection of recombinants.[46]

It has been shown that *P. aeruginosa* and other *Pseudomonas* species can express the genes of other bacteria very effectively. The bacteria for which this is true include *E. coli*,[85,86,62] *A. calcoaceticus*[87] and *M. methylotrophus*.[77] Experiments to test the generality of this expression for a range of other bacteria are currently in progress. For *M. methylotrophus*, if it is assumed that most chromosomal gene functions will be expressed in *P. aeruginosa* when transferred into that organism as the bacterial chromosomal fragment of a prime plasmid, then a mapping procedure can be proposed for *M. methylotrophus* which will not involve the isolation of mutants in the methylotroph.

The essential features of such a mapping procedure are first that a range of primes carrying *M. methylotrophus* chromosomal fragments covering a significant part of the entire chromosome can be isolated and that prime formation is not restricted to limited regions of the genome. Second, that most genes carried by those primes can be expressed in *P. aeruginosa*, the expression to be detected by complementation of *P. aeruginosa* functions not present in mutant strains of that organism. *P. aeruginosa* has features which make it particularly well suited for this task. Of prime importance is the efficient expression of other bacterial genes as mentioned above; there is a large number of mutant strains in *P. aeruginosa* PAO and there is also a chromosomal map with over 100 genes located. Finally, entry of bacterial chromosome of other genera into *P. aeruginosa* is enhanced because growth of *P. aeruginosa* at 43°C results in a reduction of native restriction endonuclease activity, enhancing the probability of prime plasmid establishment.[92]

A total of 28 primes derived from *M. methylotrophus* AS1 were characterized for their ability to complement *P. aeruginosa* PAO markers. To date, such complementation has been restricted to auxotrophic markers but there is no reason why other types of markers cannot be used. Individual primes were found to complement from 1 to 8 different loci of *P. aeruginosa* and it was possible to identify four groups of linked markers in *M. methylotrophus* AS1 comprising a total of 20 markers. By Southern hybridization, the presence of *M. methylotrophus* DNA in the primes was confirmed.

This general method of complementation mapping has advantages for those organisms such as obligate methylotrophs where mutants are hard to obtain. Provided a suitable range of mutants is available there is no need to limit the recipients of *P. aeruginosa*. The method could be used to study metabolic functions in methylotrophs, to compare such functions between various types of methylotrophs, and to examine the regulation of methylotroph genes in diverse genetic backgrounds. From the genetic point of view it has the added attraction of being a mapping procedure which does not need mutants. It is the relative location of wild-type alleles on the bacterial chromosome which is being measured.

The use of R68.45 to generate primes in facultative methylotrophs has also been reported by Hanson and Allen.[93,110] In these experiments, R68.45 was transferred to *Methylobacterium organophilum* from *E. coli* and the formation of R prime derivatives carrying the methanol dehydrogenase marker was detected using appropriate mutants of *M. organophilum* 17M.

IV. THE USES OF GENETIC INFORMATION

One of the most effective ways of using genetic data is to combine it with data obtained from other disciplines. There is a genetic approach to the solution of biological problems, just as there is a biochemical approach. A team approach using both disciplines would seem to be the most effective way for a group of microorganisms as diverse and with such experimental difficulties as the methylotrophs. The overall importance of mapping data has been referred to above, but there are other specific areas in which genetic information is effective.

A. Regulation

One such area is in the understanding of the regulation of metabolic pathways. A genetic feature of importance to regulatory mechanisms is the linkage relationships of structural and regulatory genes. A beginning to the genetic basis of regulation of C_1 metabolism has been made by O'Connor in *M. organophilum*.[31,94,95,105,106] The coordinate regulation of some or all the enzymes of the C_1 pathway was demonstrated in *M. organophilum* XX, *Hyphomicrobium* X, *Pseudomonas* 3A2, *M. ethanolicum,* and *Pseudomonas* AM1. Only in *M. organophilum* XX has it been possible to relate this data to the organization of the genes involved. Mutants of this organism were isolated which expressed pleiotropic alterations in a range of enzymes of the C_1 pathway. Four mutants were isolated which showed linkage by transformation and O'Connor and Hansen[95] and O'Connor[31] have put forward a model of the arrangement of genes coding for C_1-specific functions in *M. organophilum*. Proof of this very useful model will come from the isolation of more mutants of structural and regulatory genes and more precise mapping techniques for this organism. The present precision of mapping by transformation does not allow for precise gene ordering of such closely linked genes.

B. The Promise of Recombinant DNA Techniques

A solution to some of the problems of genetic analysis in the methylotrophs will undoubtedly come from the use of recombinant DNA technology and expression of heterologous DNA on plasmid vectors.[107] There are two main approaches. One is the analysis of the methylotroph genomes by isolating fragments of the bacterial chromosome using restriction endonucleases, cloning by means of plasmids in selected other organisms, and identifying by complementation or biochemical methods, the gene products of individual cloned chromosomal fragments. For the success of this approach the choice of vector and host are very critical. The second is the transfer of genes from nonmethylotrophic organisms to methylotrophs to modify and improve selected functions. This is an attractive scenario for the use of methylotrophs for various industrial purposes and one example will be described in detail below.

In any cloning experiments of the first type, it is important to recognize the type of question which can be asked and the limitations of the method. The general techniques of cloning have been extensively reviewed elswhere and will not be described here.[96-99] The major difficulties which can present themselves in the cloning situation where the cloned DNA is carried by an organism different to the source of the DNA include the selection of specific regions of the genome to be cloned, the expression and regulation of the gene carried by the vector plasmid and the stability of the cloned vector and its inserted DNA in the new host.

The two groups of plasmids which are likely to be most useful for DNA cloning of methylotrophs are the IncP-1 and the IncQ plasmids. The latter have been used for a variety of cloning experiments in methylotrophs, particularly when hybrids were constructed between IncQ plasmids and pBR322 to give a wider variety of restriction endonuclease sites for cloning purposes. Gautier and Bonewald[100] have cloned the gene for methanol dehydrogenase from *Pseudomonas* AM1. Fragments of *Pseudomonas* AM1 following *Eca*I digestion were ligated with *Eca*I cut R1162, an IncQ plasmid identical to RSF1010. The ligation mix was transformed into *E. coli.* The culture arising from this transformation was then conjugated with *E. coli* RP4, to mobilize the R1162 (which is transfer-deficient) into *Pseudomonas* AM1 M15a, a mutant deficient in methanol dehydrogenase and hence unable to grow on methanol. The conjugation mixture was grown overnight then plated on methanol containing agar medium which would not support growth of *E. coli* or *Ps.* AM1 M15a. Two transconjugant colonies were obtained which could grow on methanol medium although growing slower than wild-type *Pseudomonas* AM1 and these were shown to produce methanol dehydrogenase. They were shown to contain a plasmid which carried the methanol dehydrogenase gene, which could be mobilzed (by RP4) between *E. coli* and *Pseudomonas* AM1 M15a. This plasmid is unstable in *E. coli*, but can be maintained in *Pseudomonas* AM1 M15a by positive selection.

This first cloning of Am1 methanol utilizing genes illustrates many of the problems which may be encountered in experiments of this type including failure of expression of methylotroph genes in *E. coli* and the lack of a suitable selection system for such genes in nonmethylotrophic bacteria. The difficulties of transforming *Pseudomonas* AM1 with plasmid DNA were overcome to some extent by the use of RP4 (IncP-1 plasmids mobilize IncQ plasmids) but the system lacks the quantitative precision of transformation. Finally, the lack of stability of those plasmids having cloned *Pseudomonas* AM1 DNA in the *E. coli* host may have been due to the selection of the *Eca*I site, as this cloning site is close to the site of origin of DNA replication in R1162.

More recently Nunn et al.[101] have developed a more efficient cloning system for *Pseudomonas* AM1. Gene banks were prepared in a broad host range cosmid derivative of pRK290, pVK100. Again, the hybrid pVK100 cosmids had to be mobilized by another plasmid into the *Pseudomonas* AM1 mutants but in three cases complementation of the *Pseudomonas* AM1 mutants was obtained, indicating that three genes coding respectively for methanol dehydrogenase, malyl CoA lyase and hydroxypyruvate reductase had been separately cloned into the cosmid vector.

C. Genetic Modification of Methylotrophs by Recombinant DNA Technology

The most elegant use of the DNA recombinant techniques in methylotrophs has been the genetic modification of *Methylophilus methylotrophus* AS1 to increase the energy efficiency of methanol utilization. One of the reasons that *M. methylotrophus* AS1 was chosen for use in the ICI "Pruteen" process was its high carbon conversion efficiency. However, the wild-type AS1 uses the glutamine synthetase pathway for ammonia assimilation, this energetically suboptimal pathway being better for scavenging nitrogen under conditions of low ammonia concentration. Under artificial culture conditions, ammonia is not limiting, so that exchanging the native ammonia assimilation pathway of AS1 for one using glutamate dehydrogenase should result in the organism consuming less energy for this process, and hence more methanol should become available for protein synthesis and the formation of biomass.

The glutamate dehydrogenase gene *(gdh)* of *E. coli* was cloned first in an RP4::Mu derivative, then subcloned into the vector plasmid pACYC184. However this plasmid cannot be transferred to *M. methylotrophus*. The fragment of *E. coli* DNA carrying the *gdh* gene was subcloned again into pTB70, a derivative of the IncQ plasmid R300B loaded with the transposon Tn*5*. These derivatives were then mobilized to *M. methylotrophus* from *E. coli* by RP4, selecting for streptomycin resistant transconjugants. It was found that the *gdh* gene was functional in *M. methylotrophus,* although there was a reduced level of expression compared to that measured in *E. coli.* The glutamine synthetase (GS) and glutamate synthase (GOGAT) functions of *M. methylotrophus* must be blocked if the *E. coli gdh* gene is to function optimally. It was not possible to select for the appropriate type of mutant in *M. methylotrophus* but an indirect approach was successfully used. A range of temperature-sensitive mutants (demonstrating growth at 30°C, but lack of growth at 37°C) of *M. methylotrophus* were isolated then screened for growth at 37°C after they had acquired the pTB70 derivative carrying the *gdh* gene. Of 550 temperature-sensitive mutants tested four were shown to be complemented by the *E. coli gdh* gene product of the pTB70 plasmid. The enzyme activity of the *gdh* gene in *M. methylotrophus* was 10- to 20-fold lower than in *E. coli,* although the carbon conversion efficiency of the plasmid-carrying strain was 4 to 7% higher than the unmodified *M. methylotrophus.*[7]

It was subsequently shown[102] that the *M. methylotrophus* AS1 temperature-sensitive mutant carrying the pTB70 plasmid in which the *gdh* gene had been inserted was unstable in continuous culture. The *gdh* gene was retained, but the three antibiotic resistance genes of the plasmid genome were lost. Restriction endonuclease analysis showed that other regions of the plasmid were also lost. The regions of the plasmid lost were not constant as shown by the fact that plasmids of size 14.7 kb and 16.2 kb could be isolated after growth of the modified *M. methylotrophus* strains compared to the original plasmid size of 25.2 kb.

However, while this genetically modified *M. methylotrophus* strain may not have achieved all its industrial goals, it is an excellent example of the power of the recombinant DNA techniques to synthesize novel bacterial strains. Geneticists no longer have to rely solely on the genetic variability in a closely related group of organisms to construct novel strains but can call on specific functions from the entire range of microorganisms. Whereas strain improvement in the past has relied on loss of function by mutation, or rearrangement of functions by recombination of the DNA, now new functions can be acquired giving hybrids with properties which could not be achieved by any of the traditional genetic approaches.

D. The Use of Methylotrophs for the Expression of Foreign Coding Sequences

The inability of *E. coli* to express C_1 metabolism genes from methylotrophs has been referred to above. However, the reverse situation may not be true and there is industrial potential in the ability of methylotrophs to express heterologous DNA. Hennam et al.[103] have shown that eukaryotic cDNA coding for chicken ovalbumin and mouse dihydrofolate reductase are expressed well in *M. methylotrophus* AS1. The plasmid used for cloning was a hybrid of the IncQ plasmid R300B and the ColE derivative pBR322. Expression of dihydrofolate reductase in *M. methylotrophus* was higher than in *E. coli,* while the synthesis of the ovalbumin like protein was less than in *E. coli.* de Maeyer et al.[104] have also shown that a chemically synthesized human α-interferon gene on a plasmid could be expressed in *M. methylotrophus* somewhat better than in *E. coli.* As the authors point out, this organism has attractive features for the production of such pharmacological products. It grows efficiently on cheap substrates, its toxicology has been well studied in view of its use as a source of single-cell protein (Pruteen), and because it is an obligate methylotroph, it will not infect humans.

V. PROSPECT, RETROSPECT, AND SUMMARY

There seems to be general agreement in the literature and by speakers at recent conferences on C_1 metabolism that the future development of knowledge and understanding of methylotrophic bacteria is unlikely to proceed effectively without the availability of genetic techniques. It is difficult at this time to define why attempts to apply existing procedures of microbial genetics to various methylotrophs have met with less than the success expected. With the obligate methylotrophs, one obstacle has been the difficulty in obtaining mutants suitable for the selection of recombinants and this in turn has limited the development of effective systems of gene exchange. Conjugation systems in facultative methylotrophs are now being developed and can be expected to contribute effectively to the investigation of this group of organisms.

In the absence of any evidence to the contrary it should be assumed that the genomic structure of methylotrophs is basically the same as those other bacteria which have been genetically analyzed. In general terms this means a single circular double stranded chromosome, one or more plasmids, DNA repair mechanisms which are necessary to maintain a stable genome, and DNA replicating mechanisms of both chromosome and plasmids such that there is regular and equal partition of the genetic material at cell division. It will be interesting to determine whether the genome size of methylotrophs is significantly different from other, better-known bacteria.

What may be needed is a more rigorous and systematic examination of the needs of genetic analysis rather than a simplistic application of existing procedures from other bacteria to solve a range of limited problems in methylotrophs. Three main goals should be the development of effective mutagens, isolation of mutants which can be used for recombinant selection, and conjugational systems of genetic exchange. The initial goal should be the means of establishing a chromosomal map, particularly with respect to genes involved in those metabolic pathways which are concerned with C_1 metabolism.

It is unlikely that genes for C_1 metabolism in obligate methylotrophs will be other than chromosomally located. There is no example in microbial genetics to date of essential genes being located on a plasmid, such that loss of the plasmid would be a lethal event. The C_1 metabolism genes in facultative methylotrophs could be plasmid located, or located in part on a plasmid and in part chromosomally, an organizational pattern similar to that found in pseudomonads for the metabolism of a range of hydrocarbons.[48]

Recombinant DNA techniques can play an important role in identifying the genetic components of C_1 metabolism but there are a number of technical hurdles yet to be overcome. These include plasmid transformation of methylotrophs, further means of identifying specific genes from enzyme digests of methylotroph genomes, and better cloning vehicles. Problems may arise in the expression of cloned genes that have membrane-bound products. It is unlikely that cloning of individual genes of methylotrophs will solve all the problems of understanding the regulation of the C_1 pathways and the solution of that problem will need the more traditional approaches of genetic analysis.

Perhaps one aspect of the initial approach of the *E. coli* and phage geneticists should be recalled. One bacterial strain and one set of bacteriophages was selected for common study. The heterogeneity of the methylotrophs and the use of individual strains (sometimes of uncertain taxonomic status) by individual workers is leading to limited genetic knowledge for an array of different organisms. A period of a few years of common effort on a few selected microorganisms may be more effective.

This review has not attempted to assess the status of genetic knowledge of methylotrophic yeasts, nor to discuss methanogenic bacterial genetics, although it is encouraging to see articles on the genetics of these latter organisms beginning to appear.[111,112]

The methylotrophs are now prime candidates to be selected by industry for use in biotechnology. Their potential ability to convert the low cost substrate methanol into a variety of high cost end products will surely accelerate genetic interest in the methylotrophs. There is every expectation that the increased genetic knowledge of this challenging group of organisms which will result from this interest will contribute to the general understanding of the methylotrophs and more generally to microbial genetics.

ACKNOWLEDGMENTS

Research work in the author's laboratory is supported by the Australian Research Grants Scheme and Imperial Chemical Industries. The author wishes to thank those colleagues who made manuscripts in press and in preparation available to him, and Martha Sinclair, Walter Schurter and Tony Morgan for their criticisms of the manuscript.

REFERENCES

1. **Botstein, D. and Maurer, R.,** Genetic approaches to the analysis of microbial development, *Annu. Rev. Genet.*, 16, 61, 1982.
2. **Levinthal, M.,** Bacterial genetics excluding *E. coli*, *Annu. Rev. Microbiol.*, 28, 219, 1974.
3. **Holloway, B. W.,** Plasmids that mobilize bacterial chromosome, *Plasmid*, 2, 1, 1979.
4. **Riley, M. and Anilionis, A.,** Evolution of the bacterial genome, *Annu. Rev. Microbiol.*, 32, 519, 1978.
5. **Wallace, D. C. and Morowitz, H. J.,** Genome size and evolution, *Chromosoma*, 40, 121, 1973.
6. **Broda, P.,** *Plasmids*, W. H. Freeman, Oxford, 1979.
7. **Windass, J. D., Worsey, M. J., Pioli, E. M., Pioli, D., Barth, P. T., Atherton, K. T., Dart, E. C., Byrom, D., Powell, K., and Senior, P. J.,** Improved conversion of methanol to single cell protein by *Methylophilus methylotrophus*, *Nature (London)*, 287, 396, 1980.
8. **Haas, D. and Holloway, B. W.,** R factor variants with enhanced sex factor activity in *Pseudomonas aeruginosa*, *Mol. Gen. Genet.*, 144, 243, 1976.
9. **Haas, D. and Holloway, B. W.,** Chromosome mobilization by the R plasmid R68.45: a tool in *Pseudomonas* genetics, *Mol. Gen. Genet.*, 158, 229, 1978.
10. **Krishnapillai, V., Royle, P., and Lehrer, J.,** Insertions of the transposon Tn*1* into the *Pseudomonas aeruginosa* chromosome, *Genetics*, 97, 495, 1981.
11. **Weiss, A. A. and Falkow, S.,** Transposon insertion and the subsequent donor formation promoted by Tn*501* in *Bordetella pertussis*, *J. Bacteriol.*, 153, 304, 1983.
12. **Clarke, L. and Carbon, J.,** A colony bank containing synthetic ColEl hybrid plasmids representative of the entire *E. coli* genome, *Cell*, 9, 91, 1976.
13. **Williams, E., Shimmin, A., and Bainbridge, B. W.,** Mutation in the obligate methylotrophs *Methylococcus capsulatus* and *Methylomonas albus*, *FEMS Microbiol. Lett.*, 2, 293, 1977.
14. **O'Connor, M., Wopat, A., and Hanson, R. S.,** Genetic transformation, *Methylobacterium organophilum*, *J. Gen. Microbiol.*, 98, 265, 1977.
15. **Higgins, I. J., Best, D. J., Hammond, R. C., and Scott, D.,** Methane-oxidizing microorganisms, *Microbiol. Rev.*, 45, 556, 1981.
16. **Williams, E. and Shimmin, M. A.,** Radiation-induced filamentation in obligate methylotrophs, *FEMS Microbiol. Lett.*, 4, 137, 1978.
17. **Demerec, M.,** Mutagenic action of manganous chloride, *Cold Spring Harbor. Symp. Quant. Biol.*, 16, 215, 1951.
18. **Holloway, B.,** Genetic recombination in *Pseudomonas aeruginosa*, *J. Gen. Mirobiol.*, 13, 572, 1955.
19. **Kleckner, N., Roth, J., and Botstein, D.,** Genetic engineering *in vivo* using translocatable drug resistance elements: new methods in bacterial genetics, *J. Mol. Biol.*, 116, 125, 1977.
20. **Beringer, J. E., Beynon, J. L., Buchanan-Wollaston, A. V., and Johnston, A. W. B.,** Transfer of the drug-resistance transposon Tn*5* to *Rhizobium*, *Nature, (London)*, 276, 633, 1978.
21. **Lidstrom-O'Connor, M. E.,** Genetics of methylotrophy and methanotrophy, in *Microbial Chemoautotrophy*, Strohl, W. R. and Tuovinen, O. H., Eds., Ohio State University Press, Columbus, in press.
22. **Sato, M., Staskawicz, B. J., Panopoulos, N. J., Peter, S., and Honma, M.,** A host-dependent hybrid plasmid suitable as a suicidal carrier for transposable elements, *Plasmid*, 6, 325, 1981.
23. **Moore, R. J. and Krishnapillai, V.,** Tn*7* and Tn*501* insertions into the *Pseudomonas aeruginosa* plasmid R91-5: mapping of two transfer regions, *J. Bacteriol.*, 149, 276, 1982.
24. **Oki, T., Nishida, H., and Ozaki, A.,** Deoxyribonucleic acid bacteriophage of *Methanomonas methylovora*, *J. Virol.*, 9, 544, 1972.
25. **Tyutikov, F. M., Bespalova, I. A., Rebentish, B. A., Aleksandrushkina, N. N., and Krivisky, A. S.,** Bacteriophages of methanotrophic bacteria, *J. Bacteriol.*, 144, 375, 1980.
26. **Morgan, A. F.,** Transduction of *Pseudomonas aeruginosa* with a mutant bacteriophage E79, *J. Bacteriol.*, 139, 137, 1979.
27. **Wilson, G. G., Young, K. Y., and Edlin, G. J.,** High frequency generalised transduction by bacteriophage T4, *Nature (London)*, 280, 80, 1979.
28. **Drexler, H.,** Transduction by bacteriophage Tl, *Proc. Natl. Acad. Sci. U.S.A.*, 66, 1083, 1970.
29. **Williams, E. and Bainbridge, B. W.,** Genetic transformation in *Methylococcus capsulatus*, *J. Appl. Bacteriol.*, 34, 683, 1971.
30. **O'Connor, M. L. and Hanson, R. S.,** Linkage relationships between mutants of *Methylobacterium organophilum* impaired in their ability to grow on one carbon compounds, *J. Gen. Microbiol.*, 104, 105, 1978.
31. **O'Connor, M. L.,** Extension of the model concerning linkage of genes coding for C-1 related functions in *Methylobacterium organophilum*, *Appl. Environ. Microbiol.*, 41, 437, 1981.
32. **Paulchaudhuri, S. and Chakrabarty, A. M.,** Isolation of plasmid deoxyribonucleic acid from *Pseudomonas putida*, *J. Bacteriol.*, 126, 410, 1976.

33. **Warner, P. J. and Higgins, I. J.,** Examination of obligate and facultative methylotrophs for plasmid DNA, *FEMS Microbiol. Lett.*, 1, 339, 1977.
34. **Monteiro, M. J., Typas, M. A., Moffett, B. F., and Bainbridge, B. W.,** Isolation and characterization of a high molecular weight plasmid from the obligate methanol-utilizing bacterium *Methylomonas (Methanomonas) methylovora*, *FEMS Microbiol. Lett.*, 15, 235, 1982.
35. **Bainbridge, B. W. and Typas, M. A.,** personal communication.
36. **Stahl, V. and Esser, K.,** Plasmids in *Methylomonas clara*, A methylotrophic producer of single cell protein, *Eur. J. Appl. Microbiol. Biotechnol.*, 15, 223, 1982.
37. **Haber, C. L., Allen, L. N., and Hanson, R. S.,** Unpublished data quoted in Lidstrom-O'Connor, 1982.
38. **Southern, E. M.,** Detection of specific sequences among DNA fragments separated by gel electrophoresis, *J. Mol. Biol.*, 98, 503, 1975.
39. **Lidstrom-O'Connor, M. E.,** Genetics of methylotrophic bacteria, *Proc. 13th Intl. Congress Microbiology*, 1982, in press.
40. **Hanson, R. S.,** Ecology and diversity of methylotrophic organisms, *Adv. Appl. Microbiol.*, 26, 3, 1980.
41. **Warner, P. J., Higgins, I. J., and Drozd, J. W.,** Conjugative transfer of antibiotic resistance to methylotrophic bacteria, *FEMS Microbiol. Lett.*, 7, 181, 1980.
42. **Boyd, C. A. and Brammar, W. J.,** personal communication.
43. **Godfrey, A. J., Morgan, A. F., and Holloway, B. W.,** Structural instability of IncP-1 plasmids in *Pseudomonas aeruginosa* PAT involves interaction with plasmid pVS1, *J. Bacteriol.*, 144, 622, 1980.
44. **Jeyaseelan, K. and Guest, J. R.,** Transfer of antibiotic resistance to facultative methylotrophs with plasmid R68.45, *FEMS Microbiol. Lett.*, 6, 87, 1979.
45. **Tatra, P. K. and Goodwin, P. M.,** R-factor mediated chromosome mobilisation in the facultative methylotroph, *Pseudomonas* AM1, *J. Gen. Microbiol.*, 129, 2629, 1983.
46. **Holloway, B. W.,** The application of *Pseudomonad*-based genetics to methylotrophs, in *Microbial Growth on C1 Compounds*, Dalton, H., Eds., Heydon, London, 1981, 317.
47. **Sinclair, M. I., Asche, V., Morgan, A. F., and Holloway, B. W.,** Plasmid-determined tobramycin and gentamicin resistance in strains of *Pseudomonas aeruginosa* from two Sydney hospitals, *Med. J. Aust.*, 2, 283, 1981.
48. **Holloway, B. W.,** Pseudomonads, in *Genetics and Breeding of Industrial Microorganisms*, Ball, C., Ed., CRC Press, Boca Raton, in press.
49. **Bagdasarian, M., Bagdasarian, M. M., Coleman, S., and Timmis, K. N.,** New vector plasmids for cloning, in *Plasmids of Medical, Environmental and Commercial Importance*, Timmis, K. and Pühler, A., Eds., Elsevier-North Holland, Amsterdam, 1979, 411.
50. **Watson, J. M. and Holloway, B. W.,** Linkage map of *Pseudomonas aeruginosa PAT*, *J. Bacteriol.*, 136, 507, 1978.
51. **Towner, K. J.,** Chromosome mapping in *Acinetobacter calcoaceticus*, *J. Gen. Microbiol.*, 104, 175, 1978.
52. **Lacy, G. H. and Leary, J. V.,** Plasmid mediated transmission of chromosomal genes in *Pseudomonas glycinea*, *Genet. Res.*, 27, 363, 1976.
53. **Barrett, J. T., Rhodes, C. S., Ferber, D. M., Jenkins, B., Kuhl, S. A., and Ely, B.,** Construction of a genetic map for *Caulobacter crescentus*, *J. Bacteriol.*, 149, 889, 1982.
54. **Coetzee, J. N.,** Mobilization of the *Proteus mirabilis* chromosome by R plasmid R772, *J. Gen. Microbiol.*, 108, 103, 1979.
55. **Dénarié, J., Rosenberg, C., Bergeron, B., Boucher, C., Michel, M., and Barate de Bertalmis, M.,** Potential of RP4::Mu plasmids for *in vivo* genetic engineering of gram negative bacteria, in *DNA Insertion Elements Plasmids and Episomes*, Bukhari, A. I., Shapiro, J. A., and Adhya, S. L., Eds., Cold Spring Harbor Laboratory, Cold Spring Harbor, N.Y., 1977, 507.
56. **Faelen, M., Toussaint, A., von Montagu, M., van der Elsacker, S., Engler, G., and Schell, J.,** *In vivo* genetic engineering: the Mu-mediated transposition of chromosomal DNA segments into transmissible plasmids, in *DNA Insertion Elements, Plasmids and Episomes*, Bukhari, A. I., Shapiro, J. A., and Adhya, S. L., Eds., Cold Spring Harbor Laboratory, Cold Springer Harbor, N.Y., 1977, 521.
57. **Murooka, Y., Takizawa, N., and Harada, T.,** Introduction of bacteriophage Mu into bacteria of various genera and intergeneric gene transfer by RP4::Mu, *J. Bacteriol.*, 145, 358, 1981.
58. **Holloway, B. W., Krishnapillai, V., and Morgan, A. F.,** Chromosomal genetics of *Pseudomonas*, *Microbiol. Rev.*, 43, 73, 1979.
59. **Haas, D. and Riess, G.,** Spontaneous deletions of the chromosome-mobilizing plasmid R68.45 in *Pseudomonas aeruginosa* PAO, *Plasmid*, 9, 42, 1983.
60. **Willetts, N. S., Crowther, C., and Holloway, B. W.,** The insertion sequence IS*21* of R68.45 and the molecular basis for mobilization of the bacterial chromosome, *Plasmid*, 6, 30, 1981.
61. **Riess, G., Holloway, B. W., and Pühler, A.,** R68.45, a plasmid with chromosome mobilizing ability (Cma) carries a tandem duplication, *Genet. Res.*, 36, 99, 1980.

62. **Holloway, B. W., Crowther, C., Royle, P., and Nayudu, M.,** R plasmids and bacterial chromosome transfer, in *Antibiotic Resistance, Transposition and Other Mechanisms,* Mitsuhashi, S., Rosival, L., and Krcméry, V., Eds., Springer Verlag, Berlin, 1980, 19.
63. **Hamada, S. E., Luckey, J. P., and Farrand, S. K.,** R-plasmid-mediated chromosomal gene transfer in *Agrobacterium tumefasciens, J. Bacteriol.,* 139, 280, 1979.
64. **Franche, C., Canelo, E., Gauthier, D., and Elmerich, C.,** Mobilization of the chromosome of *Azospirillum brasilense* by plasmid R68.45, *FEMS Microbiol. Lett.,* 10, 199, 1981.
65. **Chatterjee, A. K. and Starr, M. P.,** Genetics of *Erwinia* species, *Annu. Rev. Micribiol.,* 34, 645, 1980.
66. **Chatterjee, A. K.,** Acceptance by *Erwinia* spp. of R plasmid R68.45 and its ability to mobilize the chromosome of *Erwinia chrysanthemi, J. Bacteriol.,* 142, 111, 1980.
67. **Nayudu, M.,** unpublished data.
68. **Beringer, J. E. and Hopwood, D. A.,** Chromosomal recombination in *Rhizobium leguminosarum, Nature (London),* 264, 291, 1976.
69. **Leonardo, J. M. and Goldberg, R. B.,** Regulation of nitrogen metabolism in glutamine auxotrophs, of *Klebsiella pneumoniae, J. Bacteriol.,* 142, 99, 1980.
70. **Kondorosi, A., Kiss, B. G., Forrai, T., Vincze, E., and Barofalvi, Z.,** Circular linkage map of *Rhizobium meliloti* chromosome, *Nature (London),* 268, 525, 1977.
71. **Megias, M., Caviedes, M. A., Palomares, A. J., and Perez-Silva, J.,** Use of plasmid R68.45 for constructing a circular linkage map of the *Rhizobium trifolii* chromosome, *J. Bacteriol.,* 149, 59, 1982.
72. **Marrs, B.,** Mobilization of the genes for photosynthesis from *Rhodopseudomonas capsulata* by a promiscuous plasmid, *J. Bacteriol.,* 146, 1003, 1981.
73. **Sistrom, W. R.,** Transfer of chromosomal genes mediated by plasmid R68.45 in *Rhodopseudomonas sphaeroides, J. Bacteriol.,* 131, 526, 1977.
74. **Skotniki, M. L., Warr, R. G., Goodman, A. E., and Rogers, P. L.,** Development of genetic techniques and strain improvement in *Zymomonas mobilis,* in *Genetics of Industrial Microorganisms 1982,* Ikeda, Y. and Beppu, T., Eds., Kodansha, Tokyo, 1983, 361.
75. **Nayudu, M. and Holloway, B. W.,** unpublished data.
76. **Dean, H. F.,** Chromosome Mapping in *Pseudomonas putida* Strain PPN, Ph.D. thesis, Monash University, Clayton, Victoria, Australia, 1982.
77. **Moore, A. T., Nayudu, M., and Holloway, B. W.,** Genetic mapping in *Methylophilus methylotrophus* AS1, *J. Gen. Microbiol.,* 129, 785, 1983.
78. **Holloway, B. W.,** Isolation and characterization of an R′ plasmid in *Pseudomonas aeruginosa, J. Bacteriol.,* 133, 1078, 1978.
79. **Morgan, A. F.,** Isolation and characterization of *Pseudomonas aeruginosa* R′ plasmids constructed by means of interspecific mating, *J. Bacteriol.,* 149, 654, 1982.
80. **Bray, R. and Morgan, A. F.,** personal communication.
81. **Johnston, A. W. B., Setchell, S. M., and Beringer, J. E.,** R primes in *Rhizobium, J. Gen. Microbiol.,* 104, 209, 1978.
82. **Kiss, G. P., Dobo, K., Dusha, I., Breznovito, A., Orosz, L., Vincze, E., and Kondorosi, A.,** Isolation and characterization of an R-prime plasmid from *Rhizobium meliloti, J. Bacteriol.,* 141, 121, 1980.
83. **Espin, G., Alvarez-Morales, A., and Merrick, M.,** Complementation analysis of *gln*-A linked mutations which affect nitrogen fixation in *Klebsiella pneumoniae, Mol. Gen. Genet.,* 184, 213, 1981.
84. **Holloway, B. W.,** unpublished data.
85. **Mergeay, M. and Gerits, J.,** F′-plasmid transfer from *Escherichia coli* to *Pseudomonas fluorescens, J. Bacteriol.,* 135, 18, 1978.
86. **Mergeay, M., Boyen, A., Legrain, C., and Glansdorff,** Expression of *Escherichia coli* K12 arginine genes in *Pseudomonas fluorescens, J. Bacteriol.,* 136, 1187, 1978.
87. **Holloway, B. W.,** *Pseudomonas* genetics and its application to other bacteria, in *Genetics of Industrial Microorganisms 1982,* Ikeda, Y. and Beppu, T., Eds., Kodansha, Tokyo, 1983, 41.
88. **Chakrabarty, A. M.,** Plasmids in *Pseudomonas, Annu. Rev. Genet.,* 10, 7, 1976.
89. **Chumley, F. G., Menzel, R., and Roth, J. R.,** Hfr formation directed by Tn*10, Genetics,* 91, 639, 1979.
90. **Johnson, S. R. and Romig, W. R.,** Transposon-facilitated recombination in *Vibrio cholerae, Mol. Gen. Genet.,* 170, 93, 1979.
91. **Haas, D., Watson, J., and Leisinger, T.,** Isolation of an Hfr donor of *Pseudomonas aeruginosa* PAO by insertion of the plasmid RP1 into the tryptophan synthase gene, *Mol. Gen. Genet.,* 182, 240, 1981.
92. **Holloway, B. W.,** Variations in restriction and modification of bacteriophage following increase of growth temperature of *Pseudomonas aeruginosa, Virology,* 25, 634, 1965.
93. **Hanson, R. S.,** personal communication.
94. **O'Connor, M. L.,** Regulation and genetics in facultative methylotrophic bacteria, *Microbial Growth on C_1 Compounds,* Dalton, H., Ed., Heydon, London, 1981, 294.

95. **O'Connor, M. L. and Hanson, R. S.,** Enzyme regulation in *Methylobacterium organophilum*, *J. Gen. Microbiol.*, 101, 327, 1977.
96. **Maniatis, T., Fritsch, E. F., and Sambrook, J.,** *Molecular Cloning — A Laboratory Manual*, Cold Spring Harbor Laboratory, Cold Spring Harbor, 1982.
97. **Timmis, K. N.,** *In vitro* methods for the manipulation of DNA, in *Genetics as a Tool in Microbiology*, Hopwood, D. A. and Glover, S. W., Eds., Cambridge University Press, Cambridge, 1982, 49.
98. **Grossman, L. and Moldave, K., Eds.,** Nucleic acids, in *Methods in Enzymology 65*, Academic Press, New York, 1980.
99. **Wu, R.,** Recombinant DNA, in *Methods in Enzymology*, Academic Press, New York, 1979, 68.
100. **Gautier, F. and Bonewald, R.,** The use of plasmid R11.62 and derivatives for gene cloning in the methanol utilising *Pseudomonas* AM1, *Mol. Gen. Genet.*, 178, 375, 1980.
101. **Nunn, D. N., Fulton, G. L., and Lidstrom, M. E.,** personal communication.
102. **Powell, K. A. and Byrom, D.,** Culture stability in strains used for single cell protein production, in *Genetics of Industrial Microorganisms*, Ikeda, Y. and Beppu, T., Eds., Kodansha, Tokyo, 1982, 345.
103. **Hennam, J. F., Cunningham, A. E., Sharpe, G. S., and Atherton, K. T.,** Expression of eukaryotic coding sequences in *Methylophilus methylotrophus*, *Nature (London)*, 297, 80, 1982.
104. **de Maeyer, E., Skup, D., Prasad, K. S. N., de Maeyer-Guignard, J., Williams, B., Meacock, P., Sharpe, G., Pioli, D., Hennam, J., Schuch, W., and Atherton, K.,** Expression of a chemically synthesized human αl interferon gene, *Proc. Natl. Acad. Sci. U.S.A.*, 79, 4256, 1982.
105. **McNerney, T. and O'Connor, M. L.,** Regulation of enzymes associated with C-1 metabolism in three facultative methylotrophs, *Appl. Environ. Microbiol.*, 40, 370, 1980.
106. **Bamforth, C. W. and O'Connor, M. L.,** The isolation of pleiotrophic mutants of *Pseudomonas aminovorans* deficient in the ability to grow on methylamine and an examination of their enzymatic constitution, *J. Gen. Microbiol.*, 110, 143, 1979.
107. **Brammar, W. J.,** Possible applications of modern genetic techniques to SCP organisms, in *Microbial Growth on C_1 Compounds*, Dalton, H., Ed., Heydon, London, 1981, 312.
108. **Harwood, J. H., Williams, E., and Bainbridge, B. W.,** Mutation of the methanic-oxidizing bacterium, *Methylococcus capsulatus*, *J. Appl. Bacteriol.*, 35, 99, 1972.
109. **Warner, P. J., Higgins, I. J., and Drozd, J. W.,** Examination of obligate and facultative methylotrophs for plasmid DNA, *FEMS Microbiol. Lett.*, 6, 339, 1977.
110. **Allen, L. N. and Hanson, R. S.,** Use of plasmid R68.45 for *in vivo* cloning of the genes for methanol utilization from *Methylobacterium organophilum*, Abstr. XIII Intl. Cong. Microbiology, Boston, 1982.
111. **Reeve, J. N., Trun, N. J., and Hamilton, P. J.,** Beginning genetics with methanogens, in *Genetic Engineering of Microorganisms for Chemicals*, Hollaender, A., Ed., Plenum Press, New York, 1982, 233.
112. **Thomm, M., Altenbuckner, J., and Stetter, K. O.,** Evidence for a plasmid in a methanogenic bacterium, *J. Bacteriol.*, 153, 1060, 1983.

Chapter 4

PRODUCTION OF METABOLITES BY METHYLOTROPHS

Yasushi Morinaga and Yoshio Hirose

TABLE OF CONTENTS

I. INTRODUCTION

In industrial fermentation processes, carbon source is one of the crucial factors for reducing cost of production. Carbohydrates of agricultural origin have been chiefly used so far, but recently unconventional carbon sources such as normal alkanes, natural gas, and carbon dioxide have been also drawing attention. Particularly, C_1 compounds are very promising feedstocks in fermentation industry. Methane, widely distributed as a main component in natural gas, is one of the most inexpensive carbon sources at present. Methanol is easily derived from methane by chemical oxidation. In recent years, the microbial production of single-cell proteins (SCP) and useful metabolites from methane and methanol has been of growing interest all over the world.

From the industrial viewpoint, methanol has several advantages over methane. The explosive property of gas mixture of methane and oxygen makes methane much less easy to handle in fermentation process. Furthermore, methane-utilizing microorganisms are limited in variety, and not able to grow so rapidly as methanol-utilizers. On the other hand, methanol serves as an efficient source of carbon and energy for a variety of microorganisms including bacteria, streptomyces, yeasts, and fungi. The assimilation of methanol requires less oxygen, and therefore evolves lower heat. Methanol of high purity is readily available at low cost. Methanol, miscible with water in all proportions, is easier to transport and store.

For the above reasons, more efforts have been paid to use methanol as a raw material for fermentation. Consequently, for example, new plants for SCP production from methanol nearly go on stream.[1-3] Today, the attention seemed to be drifting from SCP production to metabolite accumulation. Although studies are still in their infancy, characteristic properties of methylotrophs will make it possible to exploit new industrial routes to convert methanol to various useful metabolites.

II. AMINO ACIDS

Methylotrophs have specific metabolic pathways for oxidation and assimilation of C_1 compounds.[4-6] C_1 compounds are assimilated through formaldehyde, and three major pathways for its fixation are known; i.e., the serine pathway, the ribulose monophosphate (RMP) cycle, and the xylulose monophosphate (XMP) pathway. One can expect the establishment of novel processes using methylotrophs for production of peculiar metabolites which are difficult to produce by conventional fermentation processes. In the serine pathway (Figure 1), C_1-unit, 5,10-methylenetetrahydrofolate (CH_2-THF), reacts with glycine to form L-serine by the action of serine transhydroxymethylase (EC 2.1.2.1). C_1-Unit plays an indispensable role in formaldehyde fixation, therefore the serine-type methylotrophs may be characterized by a high ability of C_1-unit formation. C_1-unit relates to the formation and decomposition of several amino acids such as L-methionine, L-serine, glycine, L-tryptophan, L-histidine, nucleic acids, choline, betaine, or lignine (Figure 2). Among these metabolites, L-serine and L-methionine are regarded as the specific metabolites for the serine-type methylotrophs, because these two amino acids are synthesized by transmethylation of C_1-unit to glycine and L-homocysteine, respectively (Figure 1).

Ribulose-monophosphate (RMP) cycle is also an unique pathway found in obligate methylotrophs (Figure 3). In this pathway, formaldehyde condenses with ribulose-5-phosphate to form fructose-6-phosphate; therefore the RMP-type methylotrophs may have high ability of sugar phosphates formation. Since sugar phosphates are intermediates for formation of aromatic amino acids such as L-phenylalanine, L-tyrosine, and L-tryptophan, the RMP-type methylotrophs are fit for the production of these amino acids.

In methanol-utilizing yeasts, formaldehyde condenses with xylulose-5-phosphate to form glyceraldehyde-3-phosphate and dihydroxyacetone (Figure 4). This pathway is named as

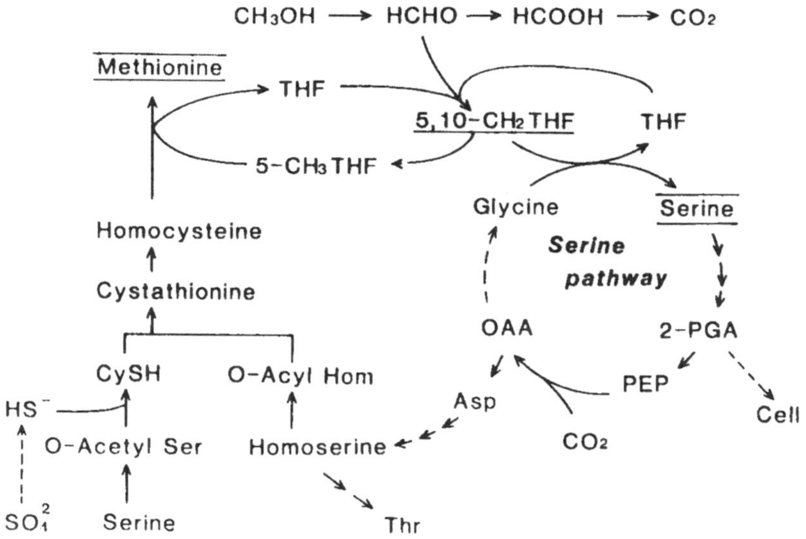

FIGURE 1. Serine pathway in methylotrophic bacteria and its relationship between biosynthetic pathways of serine and methionine.

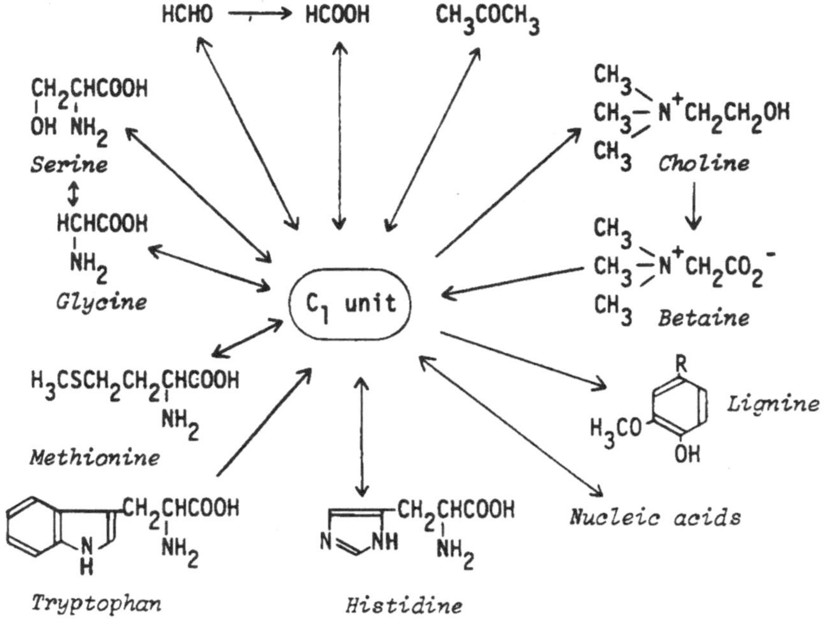

FIGURE 2. Metabolic interconversion of C_1 unit.

xylulose monosphosphate (XMP) pathway. The XMP-type methylotrophic yeasts are also suitable for the production of aromatic amino acids like the RMP-type methylotrophs.

A. Serine

L-Serine is a key intermediate for the synthesis of several other amino acids, e.g., L-

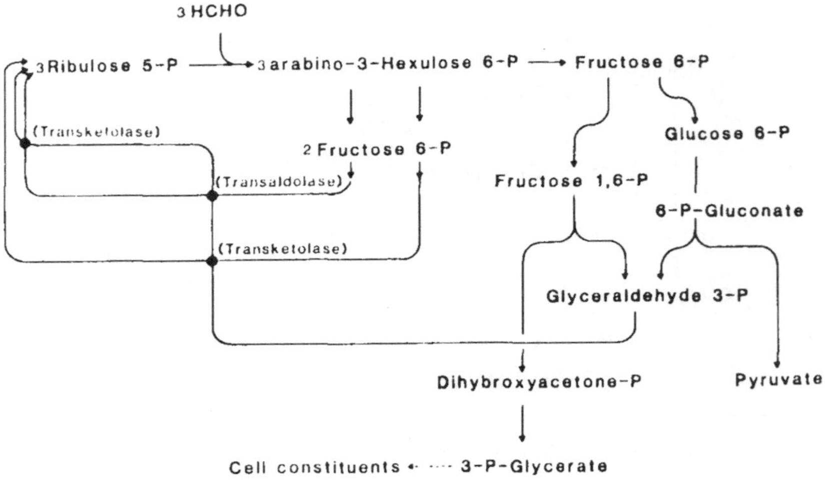

FIGURE 3. Ribulose monophosphate (RMP) pathway in methylotrophic bacteria.

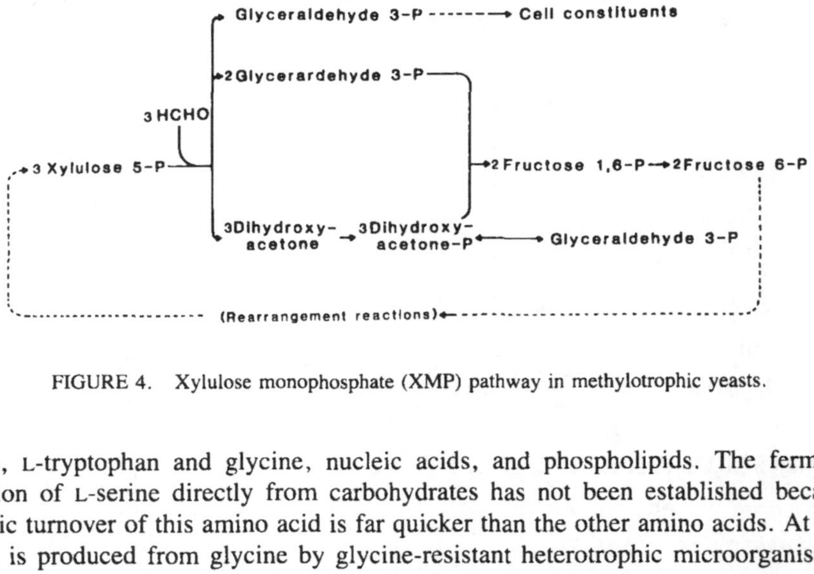

FIGURE 4. Xylulose monophosphate (XMP) pathway in methylotrophic yeasts.

cysteine, L-tryptophan and glycine, nucleic acids, and phospholipids. The fermentative production of L-serine directly from carbohydrates has not been established because the metabolic turnover of this amino acid is far quicker than the other amino acids. At present, L-serine is produced from glycine by glycine-resistant heterotrophic microorganisms such as *Corynebacterium,*[7] *Nocardia,*[8] and *Sarcina,*[9] of which growth is supported by sugars. These L-serine producers exhibit high activities of serine transhydroxymethylase (STHM) which catalyzes the condensation of C_1-unit with glycine to form L-serine. Methylotrophs are attractive as L-serine producers because their STHM has been known to be highly induced in response to methanol assimilation.

In 1976, Keune et al.[10] reported the accumulation of L-serine by a facultative serine-type methylotroph, *Pseudomonas* 3ab. When methanol was used as a sole carbon source for growth, L-serine was produced depending on the addition of glycine and methanol at the end of the exponential growth phase. Since L-serine was not derived from succinate or pyruvate as the carbon source, L-serine formation was considered to be inseparably related to methanol assimilation. The maximum yield of L-serine (4.7 g/ℓ) was obtained from 20 g/ℓ glycine and 8 g/ℓ methanol, when the pH of the culture medium was shifted to 8.5, *Pseudomonas* 3ab did not assimilate but actively degraded L-serine or glycine. The effect

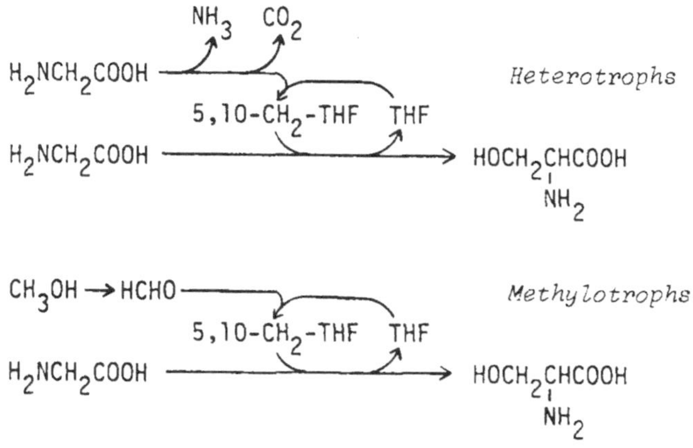

FIGURE 5. Mechanism of serine formation from glycine in serine-producing microorganisms.

of pH on L-serine accumulation was interpreted as the less degradation of L-serine and glycine in the pH range of 8.5 to 9.5.

The toxicity of glycine to the growth of microorganisms is a serious problem in fermentative production of L-serine. A serine-type methylotroph, *Arthrobacter globiformis* SK200, which was isolated by Tani et al.[11] and identified to be tolerant for glycine, accumulated up to 0.2 g/ℓ of L-serine when grown on a medium containing methanol and glycine. Although the growth was still inhibited by higher concentrations of methanol and glycine, the addition of glucose relieved the inhibition. In a combination of substrates with 1.2% methanol, 2% glycine, and 2% glucose, more than 1.5 g/ℓ of L-serine was accumulated after 4 days of cultivation. L-Methionine strongly inhibited the L-serine accumulation. This was probably due to the repression of STHM by L-methionine, as was often the case in *Escherichia coli*[12] and *Salmonella typhimurium.*[13] A methionine autotroph derived from strain SK200 accumulated 5.2 g L-serine per liter of medium containing methanol (12 g/ℓ), glycine (20 g/ℓ), and L-methionine (2 g/ℓ).

In case of L-serine production using heterotrophs,[7-9] it has been speculated that the maximum rate of conversion of glycine to L-serine is 50%, because the C_1-unit required for the synthesis of L-serine is provided by the cleavage of glycine,[14,15] and two molecules of glycine are needed to form one molecule of L-serine. In the case of methylotrophs on the other hand, one can expect up to 100% conversion of glycine to L-serine, because C_1-unit is derived from methanol (Figure 5). In this way methylotrophs are more promising than the heterotrophs.

Morinaga et al.[16-18] derived L-serine-producing mutants from a serine-type methylotroph, *Pseudomonas* MS31. The parental strain produced up to 2.5 g/ℓ of L-serine when glycine and methanol were added after its adequate growth on methanol. This strain had a disadvantageous property of decomposing L-serine actively, and the suppression of the degradation was needed for better productivity. Addition of chelating agents or divalent metal ions such as Co^{2+} and Ni^{2+} inhibited the degradation of L-serine, and increased the L-serine accumulation from glycine. A temperature-sensitive mutant, ts162, which was unable to grow on methanol above 37°C and defective in L-serine degradation, properly showed higher accumulation of L-serine (6.8 g/ℓ) from glycine (12 g/ℓ) and methanol which had been added after adequate growth of cells. In addition, the cultivation temperature was changed from 30°C (permissive) to 38 to 42°C (nonpermissive). Strain improvement was also effective for getting better productivity. The activity of STHM was known to be regulated by feedback inhibition and repression. Isozymes of STHM, one of which was a constitutive enzyme

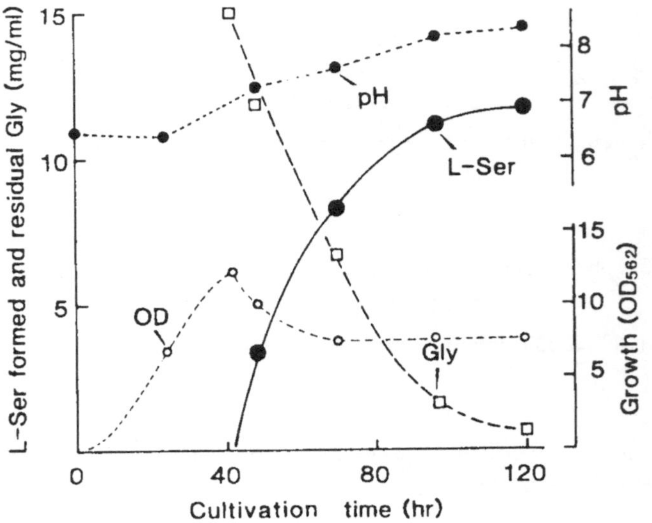

FIGURE 6. Kinetics of serine production by a mutant S395 derived from *Pseudomonas* MS31.

catalyzing the formation of glycine from L-serine and the other of which was an inducible one catalyzing the assimilation of C_1 compounds, were found in the serine-type methylotrophs, *Pseudomonas* AM1[19] and *Methylobacterium organophilum*.[20] STHM in *M. organophilum* was shown to be inhibited by glycine. In *Salmonella typhimurium*, L-serine and glycine repressed the formation of STHM. These facts implied that the derivation of the mutants resistant to analogs of L-serine or glycine was a useful way to improve the STHM activity, and to increase the L-serine productivity. Actually an *O*-methyl-DL-serine-resistant mutant, S395, derived from ts162,[18] accumulated 12 g/ℓ of L-serine from 15 g/ℓ of glycine (Figure 6). The molar conversion rate of added glycine to the product reached 57%, which was much higher than that obtained by the heterotrophic L-serine producers.

The work of Izumi et al.[22] was the highest accumulation of L-serine (24 g/ℓ) reported so far by employing the resting-cells of a new species of methylotroph, *Hyphomicorbium methylovorum* strain KM146.[23] About 17% of the molar conversion rate of added glycine to L-serine was obtained in the reaction mixture which contained methanol (24 g/ℓ), glycine (100 g/ℓ), 0.05*M* Tris-HCl (pH 9.0) and in which the cells (30 g/ℓ) were grown on methanol, with shaking at 30°C for 24 to 48 hr.

B. Methionine

Methionine is an essential amino acid which contains sulfur, and provides the sulfur and methyl groups necessary for various metabolic process. Now, racemic form of methionine is synthesized industrially from acrolein and methylmercaptan, and chiefly used for the fortification of animal feeds. However, little is known about the fermentative production of this amino acid[24,25] because it is synthesized in microorganisms via very complicated pathways. L-Homoserine, sulfide, 5-methyltetrahydrofolate and probably L-serine are indispensable intermediates in the highly branched pathways of L-methionine synthesis.

A RMP-type obligate methylotroph, OM33, has been reported to excrete 70 mg/ℓ of O-methionine when L-homoserine (5 g/ℓ) and sodium sulfide (2 g/ℓ) were added as the precursors to a methanol-containing medium.[26] A mutant of strain OM33 which was resistant to L-ethionine, an analogue of L-methionine, accumulated 420 mg/ℓ of L-methionine even in the absence of precursors.

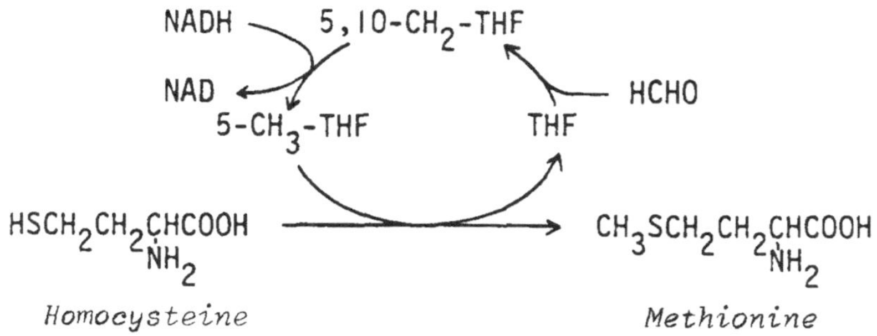

FIGURE 7. Possible mechanism for methionine formation in methylotrophs.

Since the final step of L-methionine biosynthesis is a transmethylation of C_1-unit to L-homocysteine, the serine-type methylotrophs with high activity of C_1-unit formation are promising for L-methionine production (Figure 7). Morinaga et al.[27] reported that serine-type methylotrophs actively formed L-methionine from DL-homocysteine in the presence of methanol as a methyl donor. An ethionine-resistant mutant of *Pseudomonas* FM518 having the icl^- serine pathway accumulated 0.8 g/ℓ of L-methionine in a methanol-containing medium. Supported by the elucidation of L-methionine biosynthesis,[28,29] the methylotrophs become promising microorganisms for the development of L-methionine fermentation.

C. Other Amino Acids

Oki et al.[30] reported that an obligate methylotroph, *Methanomonas methylovora,* accumulated 6.8 g/ℓ of L-glutamic acid in a yield of 11% based on methanol consumed. A strain of *Methylomonas aminofaciens* isolated by Ogata et al. produced branched chain amino acids from methanol.[31] A valine hydroxamate-resistant mutant of this strain accumulated 2.2 g/ℓ of L-valine and 0.8 g/ℓ of L-leucine.[32] RMP-type methylotrophs and XMP-type methylotrophic yeasts were found to be suitable for the production of aromatic amino acids as mentioned before. A mutant of a RMP-type methylotroph, *Methylomonas methanolophila* 6R, which had a multiple resistance to analogs of aromatic amino acids, accumulated L-phenylalanine (4 g/ℓ), L-tyrosine (1.1 g/ℓ) and L-tryptophan (0.2 g/ℓ).[33] Various derivatives of tryptophan were obtained in the cultures of 5-fluorotryptophan-, anthranilate-resistant mutants, and a phenylalanine-plus-tyrosine bradytrophic mutant bred from a XMP-type methylotrophic yeast, *Hansenula polymorpha.*[34] The overproduction of these metabolites was elucidated as due to a partial desensitization of anthranilate synthetase to feedback inhibition caused by L-tryptophan.[35] Furthermore, mutants of *H. polymorpha* were reported to accumulate tryptophan catabolites (tryptohol, indole acetic acid, and indole acetoaldehyde) extracellularly.[36] Tanaka et al.[37] reported that several mutants derived from facultative methylotrophs, *Microcyclus eburneus* and *Protaminobacter ruber,* accumulated 0.5 to 3.8 g/ℓ of *O*-methyl-L-homoserine in a medium containing methanol (2%). *O*-Methyl-L-homoserine was assumed to be formed by esterification of *O*-acetyl-L-homoserine, an intermediate of L-methionine biosynthesis, by methanol. *O*-Acetylhomoserine sulfhydrolase might catalyze this reaction, because high activity of this enzyme was recognized in several strains of serine-type facultative methylotrophs.[29]

III. VITAMINS AND COENZYMES

Vitamin B_{12} has attracted much attention in connection with the production of single-cell protein from methanol. Many facultative methylotrophs, such as strains of *Protaminobacter ruber*,[38,39] *Klebsiella* sp. No. 101,[40,41] *Pseudomonas* AM1,[42,43] *Microcyclus eburneus*,[42] pink-pigmented bacterium FM-02T,[44] and Gram-variable bacterium strain XF[45] were reported to produce vitamin B_{12}. The accumulation of the vitamin was specifically increased when the microorganisms were cultured on methanol, and the maximum yield (2.6 to 2.7 mg/ℓ) was obtained in fed-batch culture systems.[44,46] Although the yield was still lower than that of the large-scale culture of *Propionibacterium shermanii* (23 mg/ℓ) or *Streptomyces* sp. (5.7 mg/ℓ) on carbohydrate media,[48] the use of methanol is promising and attractive from a practical point of view. Vitamin B_{12} is known to be involved in such transmethylation reaction as methionine synthetase system in many organisms, but its precise role in the metabolism of methanol has not been elucidated. Recently, Ueda et al.[48] found that adenosyl-B_{12} accumulated in the culture of *Protaminobacter ruber* acted as the coenzyme in the conversion between malonyl-CoA and succinyl-CoA catalyzed by methylmalonyl-CoA mutase, and proposed a new pathway of formaldehyde fixation in which methylmalonyl-CoA mutase participated in the regeneration of glyoxylate in the icl^- serine pathway of *P. ruber*.

Coenzyme Q (CoQ) is also an attractive product of methylotrophs. CoQ homologues with different lengths of polyphenyl side chains are distributed in animals, plants, and microorganisms. CoQ_{10} is produced industrially by chemical synthesis or yeast fermentation of carbohydrates and used to treat heart diseases. Based on the CoQ type as one of the indexes, Urakami and Komagata[49] proposed a new taxonomical grouping system of the Gram-negative methanol-utilizing bacteria. Methylotrophs containing CoQ_{10} were classified in two groups: pink-pigmented facultative methanol-utilizers (group 2) and *Microcyclus* (group 3). Natori et al.[50] reported that pink-pigmented facultative methylotrophs, *Pseudomonas* N842, *Pseudomonas* AM1, *Pseudomonas* M27, and *Protaminobacter ruber* produced CoQ_{10} in cells (up to 1.52 mg/g dry cell). A mutant strain D1 obtained from *Pseudomonas* N842 after four successive mutagenic treatments accumulated six-times as much CoQ_{10} as the wild-type strain.[51] Intermittent feeding of glucose to the culture during cultivation enhanced the CoQ_{10} accumulation up to 280 mg/ℓ-broth or 13 mg/g-dry cell. Mitsubishi Gas Chemical Co., Inc., Japan, is now producing CoQ_{10} commercially using methylotrophs.[3]

Oxidation of C_1 compounds serves for regenerating high energy compounds such as NADH and ATP (Figure 8). NAD^+-linked formate dehydrogenase, which catalyzes the final step of the direct oxidation of methanol, is useful for the reduction of NAD^+.

Izumi et al.[52] reported that 50 m*M* of NAD^+ was converted to its reduced form at a conversion rate of 90% in a reaction system containing 500 m*M* formate and dried cells of a facultative methylotroph, *Arthrobacter* KM62. This regenerating system is attractive for the construction of NADH-requiring bioreactors.

Tani et al.[53] made the first report on ATP production from AMP using the oxidative phosphorylation system in the enzymatic oxidation of C_1 compounds. Zymolyase-treated cells of a methylotrophic yeast, *Kloeckera* sp. *(Candida boidinii)* No. 2201 produced ATP from AMP by a sequence of reactions including phosphorylation of AMP to ADP by adenylate kinase, oxidation of C_1 compounds to reduce NAD^+ and oxidative phosphorylation of ADP to ATP in the respiratory chain of mitochondria.

Shimizu et al.[54,55] found that the methanol-grown cells of *Kloeckera* No. 2201 contained about three times as much flavin, mainly flavin adenine dinucleotide (FAD), as those grown on other carbon sources than C_1 compounds. FAD pyrophosphorylase was derepressed and an FAD-dependent alcohol oxidase[5,6] which catalyzes oxidation of methanol, was induced during the growth of the yeast on methanol. Riboflavin or flavin mononucleotide (FMN) added to the culture of the methylotrophic yeast was converted to FAD in a yield of 45.5 mg/ℓ.[56]

IV. POLYSACCHARIDES

Methylotrophic bacteria grown on methane or methanol produced extracellular polysac-

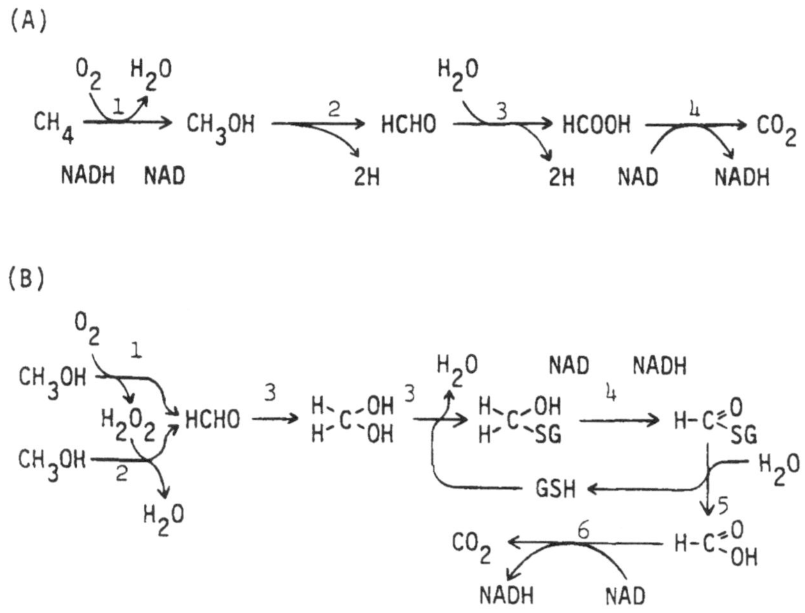

FIGURE 8. Pathway for the oxidation of methane and methanol in microorganisms. (A) Bacteria: (1) Methane monooxygenase; (2) methanol dehydrogenase; (3) formaldehyde dehydrogenase; (4) formate dehydrogenase. (B) Yeasts: (1) Alcohol oxidase; (2) catalase; (3) nonenzymatic; (4) formaldehyde dehydrogenase; (5) *S*-formylglutathione hydrolase; (6) formate dehydrogenase.

charides.[30,57-66] A newly discovered facultative methylotroph identified as *Pseudomonas oeleovorans* accumulated 400 g/ℓ of polysaccharide composed of glucose, galactose, and xylose.[62] Häggström[59] reported that a temperature-tolerant mutant of *Methylomonas methanolica* produced much more polysaccharide as its parent. Higher cultivation temperature and limited supply of growth factors and oxygen were effective for the polysaccharide accumulation. Kodama et al.[61] reported the accumulation of polysaccharide containing 76% glucose with a yield of 3.2 g/ℓ in 2% methanol-containing medium. *Pseudomonas viscogena* TS-1004 produced an acidic polysaccharide which contained D-allose (9.8%) as well as D-galactose (55.4%), D-glucose (10.7%), D-mannose (13.1%), and D-glucuronic acid (11.0%).[64] This acidic polysaccharide probably had a branched structure, and showed a specific property to form water-insoluble gel in the presence of calcium ions. An acidic polysaccharide, hyphomicran, was extracellularly produced from methanol by *Hyphomicrobium* sp. JTS-811.[65,66] Hyphomicran consisted of D-glucose, D-mannose, 2-*O*-methyl-D-mannose (a new monomethyl sugar), and pyruvic acid residues in the proportions of 2:1:1:1.[66] Hou et al.[63] reported that the methanol grown cells of *Methylocystis parvus* OBBP, which was originally isolated as a methane-utilizer, produced viscous extracellular polysaccharide which was composed of D-glucose (82%) and L-rhamnose (14%).

V. CONCLUSION

To cope with a possible worldwide protein crisis, single-cell protein (SCP) as a novel protein source has been investigated extensively. In the course of this work, a wide variety of carbon sources have been examined because the cost of the carbon substrate accounts for the major fraction of the total production cost. Initially, most of the attention was devoted to the use of hydrocarbons, particularly normal alkanes including methane. Normal alkanes were used for the production of a variety of commercially useful compounds such as amino acids and organic acids. These studies established a new research area called petroleum microbiology or petroleum fermentation. Later, methanol was of increasing interest as a carbon source for industrial fermentation because of its economic and technical advantages. Methanol was shown to be one of the most promising carbon sources for SCP production at present, and as mentioned in this chapter, production of metabolites from methanol has been extensively studied in recent years. For the production of other metabolites, however, the yields from methanol were more or less unsatisfactory for practical purposes yet. One can imagine that the difficulties in metabolite productions from methanol are that methanol is inhibitory for methylotrophs and methylotrophs found so far are still limited in variety. Continuous fed-batch culture may be promising to avoid the inhibition.[46,67-69] The exploitation of more effective methylotrophs, such as Gram-positive bacteria used for amino acid fermentation, may also be a key to the solution. Though Gram-positive bacteria have been rarely found in methylotrophs, Mimura et al.[70] recently reported the isolation of a Gram-positive one, *Corynebacterium methanolophilum* R-194. Furthermore, recombinant-DNA techniques will bring about a new evolution in this research area. Surveys for plasmids useful for gene cloning of methylotrophs are now progressing.[71-74]

REFERENCES

1. **Smith, S. R. L.,** Some aspects of ICI's single cell protein, in *Microbial Growth on C_1-compounds*, Dalton, H., Ed., Heyden, London, 1981, 342.
2. **Faust, U., Prave, P., and Schlingmann, M.,** Single cell protein from methanol; production for a high-quality product, in *Microbial Growth on C_1-compounds*, Dalton, H., Ed., Heyden, London, 1981, 335.
3. **Urakami, T., Terao, I., and Nagai, I.,** Process for producing bacterial single cell protein from methanol, in *Microbial Growth on C_1-compounds*, Dalton, H., Ed. Heyden, London, 1981, 349.
4. **Colby, J., Dalton, H., and Whittenbury, R.,** Biological and biochemical aspects of microbial growth on C_1 compounds, *Ann. Rev. Microbiol.*, 33, 481, 1979.
5. **Sahm, H.,** Metabolism of methanol by yeasts, *Adv. Biochem. Eng.*, 6, 77, 1977.
6. **Tani, Y., Kato, N., and Yamada H.,** Utilization of methanol by yeasts, *Adv. Appl. Microbiol.*, 24, 165, 1978.
7. **Kubota, K., Kageyama, K., Maeyashiki, I., Yamada, K., and Okumura, S.,** Fermentative production of L-serine, production of L-serine from glycine by *Cornebacterium glycinophilum* nov. sp., *J. Gen. Appl. Microbiol.*, 18, 365, 1972.
8. **Tanaka, Y., Araki, K., and Nakayama, K.,** Strain improvement of *Nocardia butanica* for microbial conversion of glycine into L-serine, *J. Ferment. Technol.*, 58, 163, 1980.
9. **Ema, M., Kakimoto, T., and Chibata, I.,** Production of L-serine by *Sarcina albida, Appl. Environ. Microbiol.*, 37, 1053, 1979.
10. **Keune, H., Sahm, H., and Wagner, F.,** Production of L-serine by the methanol utilizing bacterium *Pseudomonas* 3ab, *Eur. Appl. Microbiol.*, 2, 175, 1976.
11. **Tani, Y., Kanagawa, T., Hanpongkittikun, A., Ogata, K., and Yamada, H.,** Production of L-serine by a methanol-utilizing bacterium, *Arthrobacter globiformis* SK200, *Agric. Biol. Chem.*, 42, 2275, 1978.
12. **Greene, R. C. and Radovich, C.,** Role of methionine in the regulation of serine hydroxymethyltransferase in *Escherichia coli, J. Bacteriol.*, 124, 269, 1975.
13. **Stauffer, G. V. and Brenchley, J. E.,** Influence of methionine biosynthesis on serine transhydroxymethylase regulation in *Salmonella typhimurium* LT2, *J. Bacteriol.*, 129, 740, 1977.
14. **Nakazawa, E., Ajisaka, K., Kajiwara, S., and Kubota, K.,** Mechanism of L-serine formation from glycine by *Corynebacterium glycinophilum, Proc. Ann. Meeting Agric. Chem. Soc. Japan*, Yokohama, April 1 to 4, 1977, 214.
15. **Tanaka, Y., Araki, K., and Nakayama, K.,** Stimulation of microbial conversion of glycine into L-serine by magnesium phosphate, *J. Ferment. Technol.*, 58, 189, 1980.

16. **Morinaga, Y., Yamanaka, S., and Takinami, K.,** L-Serine production by methanol-utilizing bacterium *Pseudomonas* MS31, *Agric. Biol. Chem.*, 45, 1419, 1981.
17. **Morinaga, Y., Yamanaka, S., and Takinami, K.,** L-Serine production by temperature-sensitive mutants of methanol-utilizing bacterium *Pseudomonas* MS31, *Agric. Biol. Chem.*, 45, 1425, 1981.
18. **Morinaga, Y. and Hirose, Y.,** Growth of methylotrophs and its application to the production of amino acids, presented at Ann. Meet. Am. Soc. Microbiology, Atlanta, March 7 to 12, 1982.
19. **Harder, W. and Quayle, J. R.,** The biosynthesis of serine and glycine in *Pseudomonas* AM1 with special reference to growth on carbon sources other than C_1 compounds. *Biochem. J.*, 121, 753, 1971.
20. **O'Connor, M. L. and Hanson, R. S.,** Serine transhydroxy-methylase isozymes from a facultative methylotroph, *J. Bacteriol.*, 124, 985, 1975.
21. **Stauffer, G. V., Baker, C. A., and Brenchley, J. E.,** Regulation of serine transhydroxymethylase activity in *Salmonella typhimurium, J. Bacteriol.*, 120, 1017, 1974.
22. **Izumi, Y., Takizawa, M., Tani, Y., and Yamada, H.,** L-Serine production by resting cells of a methanol-utilizing bacterium, *J. Ferment. Technol.*, 60, 269, 1982.
23. **Izumi, Y., Takizawa, M., Tani, Y., and Yamada, H.,** An obligate methylotrophic *Hyphomicrobium* strain. Identification, growth characteristics and cell composition, *J. Ferment. Technol.*, 60, 371, 1982.
24. **Dulaney, E. L., Jones, C. A., and Dulaney, D. D.,** Amino acid accumulation, principally alanine, by auxotrophs of *Ustilago maydis, Dev. Ind. Microbiol.*, 5, 242, 1964.
25. **Kase, H. and Nakayama, K.,** L-Methionine production by methionine analog resistant mutants of *Cornebacterium glutamicum, Agric. Biol. Chem.*, 39, 153, 1975.
26. **Yamada, H., Morinaga, Y., and Tani, Y.,** L-Methionine over-production by ethionine-resistant mutants of obligate methylotroph strain OM33, *Agric. Biol. Chem.*, 46, 47, 1982.
27. **Morinaga, Y., Tani, Y., and Yamada, H.,** L-Methionine production by ethionine-resistant mutants of facultative methylotroph, *Pseudomonas* FM518, *Agric. Biol. Chem.*, 46, 473, 1982.
28. **Morinaga, Y., Tani, Y., and Yamada, H.,** Regulatory properties of L-methionine biosynthesis in obligate methylotroph OM33: role of homoserine-*O*-transsuccinylase, *Agric. Biol. Chem.*, 46, 57, 1982.
29. **Morinaga, Y., Tani, Y., and Yamada, H.,** Biosynthesis of homocystein in a facultative methylotroph, Pseudomonas FM518, *Agric. Biol. Chem.*, 47, 2855, 1983.
30. **Oki, T., Kitai, A., Kouno, K., and Ozaki, A.,** Production of L-glutamic acid by methanol-utilizing bacteria. *J. Gen. Appl. Microbiol.*, 19, 79, 1973.
31. **Ogata, K., Izumi, Y., Kawamori, M., Asano, Y. and Tani, Y.,** Amino acid formation by methanol-utilizing bacteria, *J. Ferment. Technol.*, 55, 444, 1977.
32. **Izumi, Y., Asano, Y., Tani, Y., and Ogata, K.,** Formation of valine and leucine by analog-resistant mutants of an obligate methylotroph, *Methylomonas aminofaciens, J. Ferment. Technol.*, 55, 452, 1977.
33. **Suzuki, M., Berglund, A., Unden, A., and Heden, C. G.,** Aromatic amino acids production by analogue-resistant mutants of *Methylomonas methanolophila* 6R, *J. Ferment. Technol.*, 55, 466, 1977.
34. **Denenu, E. O. and Demain, A. L.,** Derivation of aromatic amino acid mutants from a methanol-utilizing yeast, *Hansenula polymorpha, Appl. Environ. Microbiol.*, 41, 1088, 1981.
35. **Denenu, E. O. and Demain, A. L.,** Enzymatic basis for overproduction of tryptophan and its metabolites in *Hansenula polymorpha* mutants, *Appl. Environ. Microbiol.*, 42, 497, 1981.
36. **Denenu, E. O. and Demain, A. L.,** Relationship between genetic deregulation of *Hansenula polymorpha* and production of tryptophan metabolites, *Eur. J. Appl. Microbiol. Biotchnol.*, 13, 202, 1981.
37. **Tanaka, Y., Araki, K., and Nakayama, K.,** Accumulation of *O*-methyl-L-homoserine by methanol-utilizing bacteria, *Biotechnol. Lett.*, 2, 67, 1980.
38. **Tanaka, A., Ohya, Y., Shimizu, S., and Fukui, S.,** Production of vitamin B_{12} by methanol-assimilating bacteria, *J. Ferment. Technol.*, 52, 921, 1974.
39. **Sato, K., Ueda, S., and Shimizu, S.,** Form of vitamin B_{12} and its role in a methanol-utilizing bacterium, *Protaminobacter ruber, Appl. Environ. Microbiol.*, 33, 515, 1977.
40. **Nishio, N., Yano, T., and Kamikubo, T.,** Isolation of methanol-utilizing bacteria and its vitamin B_{12} production, *Agric. Biol. Chem.*, 39, 21, 1975.
41. **Nishio, N., Yano, T., and Kamikubo, T.,** Further studies of vitamin B_{12} production by methanol-utilizing bacterium, *Klebsiella* sp. No. 101, *Agric. Biol. Chem.*, 39, 207, 1975.
42. **Nishio, N., Tanaka, M., Matsumoto, R., and Kamikubo, T.,** Production of vitamin B_{12} by methanol-utilizing bacteria *Pseudomonas* AM1 and *Microcyclus eburneus, J. Ferment. Technol.*, 55, 200, 1977.
43. **Tsuchiya, Y., Nishio, N., and Nagai, S.,** Vitamin B_{12} from methanol by continuous culture of *Pseudomonas* AM1, *J. Ferment. Technol.*, 58, 485, 1980.
44. **Toraya, T., Yongsmith, B., Tanaka, A., and Fukui, S.,** Vitamin B_{12} production by a methanol-utilizing bacterium, *Appl. Microbiol.*, 30, 477, 1975
45. **Dumenil, G., Cremieux, A., Chevalier, J., and Guiraud, H.,** Vitamin B_{12} formation by a gram-variable methanol-utilizing bacterium, *Biotechnol. Lett.*, 3, 285, 1981.
46. **Yano, T., Kobayashi, T., and Shimizu, S.,** Fed-batch culture of methanol-utilizing bacterium with DO-stat, *J. Ferment. Technol.*, 56, 416, 1978.

47. **Wuest, H. M. and Perlman, D.,** Vitamin B_{12}. III. Industrial preparation and production, in *The Vitamins,* Vol. 2, Sebrell, W. H., Jr. and Harris, R. S., Eds., Academic Press, New York, 1968, 139.
48. **Ueda, S., Sato, K., and Simizu, S.,** Glyoxylate formation from mesaconyl-CoA and its related reactions in a methanol-utilizing bacterium, *Protaminobacter ruber, Agric. Biol. Chem.,* 45, 823, 1981.
49. **Urakami, T. and Komagata, K.,** Cellular fatty acid composition and coenzyme Q system in Gram-negative methanol utilizing bacteria, *J. Gen. Appl. Microbiol.,* 25, 343, 1979.
50. **Natori, Y., Nagasaki, T., Kobayashi, A., and Fukawa, H.,** Production of coenzyme Q_{10} by *Pseudomonas* N842, *Agric. Biol. Chem.,* 42, 1799, 1978.
51. **Natori, Y. and Nagasaki, T.,** Enhancement of coenzyme Q_{10} accumulation by mutation and effects of medium components on the formation of coenzyme Q homologs by *Pseudomonas* N842 and mutants, *Agric. Biol. Chem.,* 45, 2175, 1981.
52. **Izumi, Y., Mishra, S. K., Ghosh, B. S., Tani, Y., and Yamada, H.,** Production of NADH from NAD by resting-cell reaction with a methanol-utilizing bacterium containing high level of formate dehydrogenase, *Proc. Annu. Meet. Agric. Chem. Soc. Japan,* Tokyo, April 1 to 4, 1982, 185.
53. **Tani, Y., Mitani, Y., and Yamada, H.,** Utilization of C_1 compounds: Phosphorylation of adenylate by oxidative phosphorylation in *Candida boidinii* (*Kloeckera* sp.) No. 2201, *Agric. Biol. Chem.,* 46, 1097, 1982.
54. **Shimizu, S., Ishida, M., Tani, Y., and Ogata, K.,** Flavin changes of *Kloeckera* sp. No. 2201 during adaptation to methanol, *Agric. Biol. Chem.,* 41, 423, 1977.
55. **Shimizu, S., Ishida, M., Kato, N., Tani, Y., and Ogata, K.,** Derepression of FAD pyrophosphorylase and flavin changes during growth of *Kloeckera* sp. No. 2201 on methanol, *Agric. Bol. Chem.,* 41, 2215, 1977.
56. **Shimizu, S., Ishida, M., Tani, Y., and Ogata, K.,** Production of flavin adenine dinucleotide by methanol-utilizing yeasts, *J. Ferment. Technol.,* 55, 630, 1977.
57. **Chalfan, Y. and Mateles, R. I.,** New pseudomonad utilizing methanol for growth, *Appl. Microbiol.,* 23, 135, 1972.
58. **Dahl, J. S., Mehta, R. J., and Hoare, D. S.,** New obligate methylotroph, *J. Bacteriol.,* 109, 916, 1972.
59. **Häggström, L.,** Mutant of *Methylomonas methanolica* and its characterization with respect to biomass production from methanol, *Appl. Environ. Microbiol.,* 33, 567, 1977.
60. **Tam, K. T. and Finn, R. K.,** Extracellular microbial polysaccharides, *ACS Symp.,* 45, 58, 1977.
61. **Kodama, T., Nakahara, T., Ohmori, T., Binh, N. T., Hoshino, K., and Minoda, Y.,** Production of extracellular polysaccharides by a hydrogen bacterium and a methanol-utilizing bacterium, *Proc. 2nd Int. Symp. Microbial Growth C_1 Compounds,* Puschino, U.S.S.R., 1977, 211.
62. **Egorov, A. M., Platonenkova, L. S., Zakharova, E. V., and Netrusov, A. I.,** New strains of methylotrophic bacteria *Pseudomonas oleovolans,* a polysaccharide producer, *Prikl. Biokhim. Microbiol.,* 12, 524, 1976; *Chem. Abstr.,* 85, 90069m, 1976.
63. **Hou, C. T., Laskin, A. I., and Patel, R. N.,** Growth and polysaccharide production by *Methylocystis parvus* OBBP on methanol, *Appl. Environ. Microbiol.,* 37, 800, 1978.
64. **Misaki, A., Tsumuraya, Y., Kakuta, M., Takemoto, H., and Igarashi, T.,** D-Allose-containing polysaccharide synthesized from methanol by *Pseudomonas* sp., *Carbohydrate Research,* 75, C8, 1979.
65. **Kawamura, K., Hieda, T., Iwamuro, Y., Mikami, Y., Obi, Y., and Kisaki, T.,** Isolation and characterization of a *Hyphomicrobium* species and its polysaccharide formation from methanol, *Agric. Biol. Chem.,* 46, 2411, 1982.
66. **Kawamura, K., Iwamuro, Y., Mikami, Y., Obi, Y., and Kisaki, T.,** 2-*O*-Methyl-D-mannose in an extracellular polysaccharide from *Hypohmicrobium* sp., *Agric. Biol. Chem.,* 46, 2419, 1982.
67. **Reuss, M., Gnieser, J., Reng, H. G., and Wanger, F.,** Extended culture of *Candida boidinii* on methanol, *Eur. J. Appl. Microbiol.,* 1, 295, 1975.
68. **Yamane, T., Kishimoto, M., and Yoshida, F.,** Semi-batch culture of methanol-assimilating bacteria with exponentially increased methanol feed, *J. Ferment. Technol.,* 54, 229, 1976.
69. **Nishio, N., Tsuchiya, Y., Hayashi, M., and Nagai, S.,** A fed batch culture of methanol-utilizing bacteria with pH stat, *J. Ferment. Technol.,* 55, 151, 1977.
70. **Mimura, A. Wada, M., Sakashita, H., Hashimoto, Y., and Sowa, T.,** Isolation and characterization of a Gram-positive methanol assimilating bacterium, *J. Ferment. Technol.,* 56, 243, 1978.
71. **Gautier, F. and Bonewald, R.,** The use of plasmid R1162 and derivatives for gene cloning in the methanol-utilizing *Pseudomonas* AM1, *Mol. Gen. Genet.,* 178, 375, 1980.
72. **Windass, J. D., Worsey, M. J., Pioli, E. M., Pioli, D., Barth, P. T., Atherton, K. T., Dart, E. C., Byrom, D., Powell, K., and Senior, P. J.,** Improved conversion of methanol to single cell protein by *Methylophilus methylotrophus, Nature (London),* 287, 396, 1980.
73. **Hennam, J. F., Cunningham, A. E., Sharpe, G. S., and Atherton, K. T.,** Expression of eukaryotic coding sequence in *Methylophilus methylotropus, Nature (London),* 297, 80, 1982.
74. **Stahl, U. and Esser, K.,** Plasmids in *Methylomonas clara,* a methylotrophic producer of single cell protein, *Eur. J. Appl. Microbiol. Biotechnol.,* 15, 223, 1982.

Chapter 5

SINGLE-CELL PROTEIN

K. A. Powell and B. L. F. Rodgers

TABLE OF CONTENTS

I. INTRODUCTION

Many reviews and descriptions of single-cell protein (SCP) have been published.[1-4] The name SCP was first used by workers at MIT in 1968. It means in almost all cases the production of a food or feedstuff from single celled organisms (yeast or bacteria) but has been more loosely applied to fungal products. This chapter is not another review of biochemistry, genetics, or microbiology but an attempt to place the results of research effort in these disciplines in the context of the development of the process. In the case of the ICI process begun in 1969 much of the process development work occurred simultaneously with the development of the scientific background; this was one of the reasons for the excitement of working on such a process.

It should not be forgotten that microbiological science is only a small part of the process; SCP has presented challenges to chemical engineers, physicists, mathematicians, toxicologists, nutritionists, agriculturalists, and specialists in many other disciplines; the scale and continuous nature of the process demand that it should be so.

We will therefore attempt to describe how an industrial organization has approached the development of an SCP process, from the choice of substrate to the commissioning of the world's largest continuous fermentation plant.

II. CHOICE OF SUBSTRATE, ORGANISM, AND REACTOR

A. The Choice of Substrate

Waste products — Many SCP processes have been suggested based on the use of waste products. Few waste products are available in the right quantity and in a clean, sufficiently pure form to be usable to provide good quality feedstuff, however, the Pekkilo process has been successfully developed based on sulfite liquor from paper pulp manufacture.[6]

Carbohydrates — Carbohydrate SCP processes are rarely discussed, however there is merit in the use of such substrates where the carbohydrate is available, where protein is in short supply, and where there is sufficiently developed farming practice to use the feedstuff.

Chemical process feedstocks: alkanes — Two main feedstocks have been considered, first the alkanes, a low value fraction from oil. BP among others developed the use of alkanes for single-cell protein from yeast. The process has been largely abandoned although plants were built to produce the product.[7] Probably the largest factor was the changing value of the substrate in the late 1970s.

Methane as substrate — The second group of substrates are methane and methanol. ICI, Shell, and BP all considered SCP processes based on methane. ICI decided not to pursue a methane process for three reasons, first and most important, methane gives mass transfer problems when used as a substrate in an aqueous process; explosion hazards and the lack of a suitable organism gave other difficulties; hence ICI decided to look at another substrate — methanol.

Methanol as substrate — ICI is a world leader in methanol technology; it produces methanol in-house from North Sea gas and also provides a major share of catalysts used to make methanol throughout the world. Obviously then it made sense for ICI to consider methanol as a substrate. Methanol gives no mass transfer problems, it does not give explosive mixtures in the fermentor and microorganisms grow on it in pure culture with good yield; hence methanol was chosen and the next phase of the project commenced — the choice of organism.

B. Selecting a Process Organism

An organism for a methanol SCP process has to be chosen to comply with a number of criteria:

1. It must give a high yield on methanol as sole carbon and energy source.
2. It must not require any added growth factors since these are expensive and may cost more than the rest of the process.
3. It must grow at a temperature which will permit adequate temperature control of the fermentor in most climates.
4. It must be nonpathogenic.
5. It must yield a nutritious and nontoxic product.
6. It must be robust, so that it can withstand the rigors of large-scale cultivation.

When considering the development of a large-scale process it would be ideal to have an organism available which would pass most of these tests, but in the case of methanol SCP in the early 1970s no such organism existed; even now few organisms have been shown to comply with all the requirements. ICI microbiologists began their search for the appropriate organism by traditional mechanisms: first samples of soil, water, vegetation, etc., were taken, inoculated into a methanol salts medium and incubated at a variety of temperatures. Organisms were found which grew well in the medium; they were then purified by streaking onto methanol minimal agar. A variety of organisms were isolated; those which grew quickly to a high density were selected for further screening.

The benefit of hindsight enables us now to see the success of this screening policy. Although many methylotrophs were isolated, those which were selected for further study by the empirical screen were almost all obligate methylotrophs and almost certainly use the ribulose monophosphate pathway of formaldehyde incorporation. Facultative methylotrophs which used the serine pathway were also selected but slower growing methylotrophs *Hyphomicrobia,* etc., were all rejected.

Having studied as many parameters as possible in shake-flask culture the next step was to use 5-ℓ continuous fermentors to examine the best strains for scale-up. Figure 1 is an example of the type of reactor used for this process. A minimal medium was used with pH control provided by addition of ammonia which was also used as nitrogen source. Methanol was fed separately as the limited nutrient.

The organism which gave the best overall performance in these trials was called *Methylophilus methylotrophus* and has now been developed as the organism of choice for SCP production.

If one looks at the other commercial companies which have studied production of single-cell protein (SCP) from methanol and their methods of organism isolation it becomes clear that the route followed by ICI to choose a process organism has also been followed by others. Three companies have worked on methanol SCP to small pilot plant scale; Hoechst, Mitsubishi Gas Chemical, and Phillips Petroleum have also looked at methanol SCP. All these companies have studied methylotrophs (given various names but essentially similar to *M. methylotrophus*). All the organisms which reached pilot plant scale were obligate methylotrophic bacteria using the ribulose monophosphate pathway of formaldehyde fixation. A commercial process using yeast grown on methanol has not been developed although Phillips Petroleum, Mitsubishi Gas Chemical, and the Institute Francais du Petrole designed such a process — it is worthwhile to examine the problems and advantages associated with yeast production.

There are obvious points of advantage; yeasts have been "traditional" foodstuffs; hence they might be more acceptable (this seems a weak argument but is often used[8]). Yeast cells are large and hence more easily recovered. Yeasts can be grown to have low nucleic acid content. The problems of methanol growth yeast cultures are caused by the biochemical pathways for methanol utilization found in yeast.

The crucial point is the first step of methanol oxidation in the yeasts a flavoprotein oxidase is responsible for the oxidation of methanol. This oxidase has a low affinity for methanol

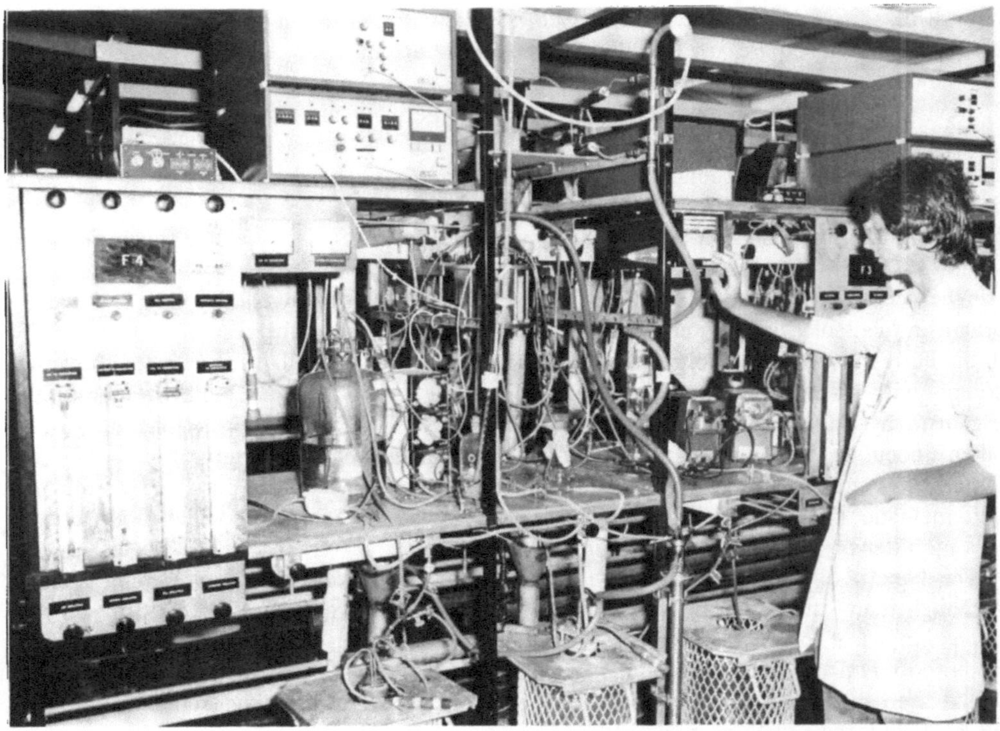

FIGURE 1. A 5 l Fermentor — the first increase in scale.

and oxygen[9] and provides no reductant or ATP for biosynthetic reactions; in short, it is extremely inefficient. The bacteria have a methanol dehydrogenase which provides some ATP. The consequences of this difference are as follows:

1. The yield of yeast is lower (by as much as 25%) than the yield of bacteria per gram of methanol.[10]
2. The yield of yeast is lower than the yield of bacteria per mole of oxygen.
3. Yeast cultures may easily become oxygen limited because of the low affinity of the enzyme for oxygen.
4. High methanol concentrations are found in yeast cultures, again because of the low affinity of the enzyme for methanol.
5. The low overall efficiency of the yeast reaction leads to a requirement for extra cooling in the fermentor.

Thus an academically interesting difference in methanol oxidation becomes of crucial importance to large-scale fermentation.

C. Choice of Reactor

SCP offers a new challenge to the process engineer of fermentation at enormously increased scale. Plants of capacities of 250,000 t/annum have been suggested. The ICI plant can produce in excess of 50,000 t/annum. It becomes obvious that fermentors much larger than those used in the traditional fermentation industry are required for such production. The scale of production raises new problems of type and number of fermentors to be used, also in how to dry and recover water from the cells produced. All these problems had to be solved in order to design a commercial plant. Few of the problems are related directly to microbiology but the properties of the organism are linked to the solutions found. We shall now go on to attempt to explain the approaches taken.

In the early 1970s the fermentation industry relied heavily on stirred tank reactors. These reactors give several problems when used on a large scale. The stirring means that there has to be a breach in the sterile boundary of the fermentor to allow access of the stirrer shaft, hence problems are caused in achieving sterility over long continuous operation.

Large-scale fermentation presents new problems in obtaining homogeneous environments within the fermentor. For these reasons ICI chose to attempt a novel solution, the "pressure cycle fermentor".

We shall now go on to describe the problems involved in scaling up this fermentor to the full scale of production. We have chosen to do this rather than to describe the historical path through pilot plant to main plant design. The operation of the pilot plant has been described elsewhere.[11] As ICI is the only company with full-scale production experience of protein from methanol this should be the most useful approach.

III. PROCESS CONSIDERATIONS AND OPTIONS

A. Introduction

The objective is to design a large-scale SCP plant which utilizes a methanol substrate. In order to achieve the objective it is necessary to answer a series of questions. The first questions in the series are fundamental and strategic, such as where should the plant be located. What should be the plant capacity and how much will it cost (roughly)? In seeking answers to these preliminary questions various resources need to be consulted, especially marketing and accounting.

Having answered the initial questions it is found that subsequently more and more detailed technical issues arise. For instance, what type of drying stage is required to give the correct product form and what is the required inlet drier gas temperature. To obtain suitable answers now, the resources to be consulted are engineering, process design, scientific, and R & D.

The following sections are arranged as a series of questions and discussions in order to assist the reader in understanding not only the questions but also the order in which they might occur in a real design exercise. However, it is frequently not possible to preserve this linear approach, tackling one problem after another. Often it is found that the answer to an issue yet to be addressed is required to solve the current problem. Clearly then an iterative approach is necessary. An example of this is found in the fermentor design section. Starting at the question of what oxygen transfer efficiency will be achieved, it is found that the answer is dependent upon the outcome of a decision on the fermentor dimensions. Indeed, the choice of fermentor dimensions will depend upon the decision about how much heat must be removed from the fermentation.

Section II covers the overall mass balance considerations allowing the scale of the plant and its components to be assessed and visualized.

Section III deals specifically with the more important aspects of plant design. The section is divided up into unit operations which together comprise the total plant.

B. Overall Mass Balance Considerations

What plant capacity is required? An overall view of the process may be obtained by answering this question. For instance, say, desired plant output is 200,000 t/annum of SCP. Assuming a product moisture content of 8% w/w and an extracellular ash and soluble organics inclusion of 5% w/w, then the following calculation applies:

Fermentor output (assuming no downstream process losses)	$= 200{,}000 \times (1.0 - 0.08 - 0.05)$
	$= 174{,}000$ dry cells/annum

Now, an assumption needs to be made about the plant overall availability. This is usually expressed in terms of a percent time, and for plants with solids handling equipment a typical availability figure is 85%. Since the SCP plant contains driers, this constitutes solids handling; hence the 85% availability figure is justifiable. It is unlikely that the fermentation area will have a lower availability since the type of hardware involved is reliable (pumps, compressors, and steam systems). If redundancy is built in to, say, the drier section, the availability obviously can be improved but at the penalty of the cost incurred in installing the spare equipment.

Using the 85% figure, and adding a figure of 95% for the overall efficiency of product recovery from the fermentor to the finished product section, the calculation continues:

$$\text{Fermentor actual output required} = \frac{1}{0.95} \times \frac{174{,}000}{365 \times 0.85} \times 24 \text{ t cells/hr}$$

$$= 24.6 \text{ t cells/hr}$$

At this stage the data produced by R&D must be consulted. The decision has to be taken as to what growth rate and cell concentration is desired in the fermentor.

Generally speaking, it is always desirable to maximize the product of the growth rate and the cell concentration since this product represents a volumetric productivity as can be seen from the dimensional considerations below:

D = Dilution rate — dimension (hr^{-1})

μ = growth rate (= dilution rate at steady state) — dimension (hr^{-1})

χ = cell concentration — dimension (kgm^{-3})

$$\text{Productivity} = \mu\chi = D\chi \ (\text{kg m}^{-3}\ \text{hr}^{-1})$$

By maximizing productivity, optimum use is made of the total working volume inside the fermentor vessel.

The allowable vaues of μ and χ are assessed from the R&D data available. This data will state what range of dilution rates or growth rates have been found to be satisfactory in respect of final cellular composition. For instance, the ribonucleic acid content of cells is associated with the production and transport of proteins within the cell. It is found that the ribonucleic acid (RNA) level in cells increases with increasing growth rate since cells growing at high growth rates tend to have a higher protein content, thus requiring more RNA.

If a cellular makeup with a high RNA content is desired, the growth rate of the cells must therefore be high (typically 0.15 to 0.2 hr^{-1}). Similarly, fats, carbohydrates, and ash levels within the cell vary with growth rate.

Having fixed the growth rate at the optimum level, the cell concentration then has to be determined. At the optimum growth rate, the maximum cell concentration that can be achieved will be that which exists just before the culture becomes limited by oxygen transfer rather than by carbon substrate concentration (methanol). In other words, oxygen transfer capability determines cell concentration for a given growth rate.

In the laboratory scale fermentor in continuous operation, the oxygen transfer depends upon the fermentor overpressure, the energy input to the culture and the way the energy is expended, the medium oxygen transfer coefficient, and the partial pressure of O_2 in the oxygen supply gas. Therefore, equally, these variables will affect the O_2 transfer in a large scale fermentor. They will also affect the achievable maximum cell concentration in the large scale. However, it is highly unlikely that in going from small- to large-scale fermentations, the same ability to transfer oxygen can be achieved since the variables mentioned above also change with scale-up. How, therefore, does one pick the cell concentration that is applicable on the large scale for this initial mass balance exercise?

The safest way to proceed is to use the cell concentration achieved consistently in the laboratory while producing an acceptable product. The aeration rate (usually expressed as vvm) should be noted and later on in the design exercise the design oxygen transfer capability of the large-scale fermentor should be checked against that actually achieved in the laboratory. Typical cell concentrations are in the range 25 to 35 (kg/m^3) where m^3 refers to fermentor vessel ungassed working volume.

So typically the productivity expected will be in the range:

$$3.5 \text{ to } 7 \ (\text{kg m}^{-3}\ \text{hr}^{-1})$$

Now the desired output of cells is 24,600 kg/hr. So the required ungassed fermentor working volume is in the range:

$$7028 \text{ to } 3514 \text{ m}^3$$

Picking a figure in this range (for the sake of simplicity in further calculations) at say 5000 m^3, the harvesting rate of flow from the fermentor may be computed (assuming a chosen dilution rate of 0.15 hr^{-1})

$$D = 5000 \times 0.15 = 750 \ (\text{m}^3\ \text{hr}^{-1})$$

Although a fermentor ungassed working volume has been calculated, it is essential to have a view of extra capacity necessary to allow for gas holdup and gas disengagement. For the former, gas holdup in the range 0 to perhaps 45% may have to be catered for depending upon the surfactants concentration within the fermentor. So the maximum gassed working volume will be

$$\text{Max. gassed working volume} = \frac{5000}{0.55} = 9090 \text{ m}^3$$

Gas disengagement is achieved at the top of the fermentor by certain configurations of vessel layout or by providing a large open surface area. An extra 30% volume should be allowed for disengagement, taking the total fermentor volume required to 11,800 m^3. It is highly unlikely that such a fermentation volume could be achieved in only one vessel.

What will the downstream stages look like? Turning now to the dewatering requirement on the fermentor harvesting stream, the flow rate is 750 (m^3/hr) ungassed and since the desired production rate is 24.6 t/hr the cell concentration needs to be 32.8 kg/m^3. So to achieve an 8% w/w moisture product:

$$\begin{array}{l}\text{Amount of moisture} \\ \text{to be removed}\end{array} = \left(\frac{24.6}{0.0328} - 24.6\right) - \left(\frac{24.6}{0.92} - 24.6\right)$$

$$= 24.6\ [(29.49) - (0.087)] = 723 \text{ t/hr}$$

(Assuming that both solid and liquid have an SG of 1.0.) Expressing this per unit cell production:

$$\begin{matrix}\text{Amount of moisture} \\ \text{to be removed/t cells}\end{matrix} = \frac{723}{24.6} = 29.4 \text{ t/t}$$

Note that in this calculation solubles have been ignored for the time being.

Using a figure of £3.5/t water evaporated as a rough guide to the cost of single stage drying by evaporation (1982 prices), then the drying cost per tonne of cells would be almost £105/t cells. Clearly it would not be economic to propose a process such as this. Water needs to be removed by less expensive means as much as possible before resorting to expensive thermal means.

The most elegant solution to the dewatering problem is achieved by making the cells flocculate and float out of the culture within which they were originally suspended. The flotation process achieves its effect by virtue of bubbles of dissolved gases being formed when the pressure imposed on the culture is released. The process only works when intracellular "cements" such as nucleic acids and proteins are released into the culture, effectively binding together cellular debris and particulates around forming gas bubbles. The flocculated particles and gas bubbles then rise to the surface of the culture to produce a blanket of highly concentrated solids. To release intracellular cements it is necessary to thermally shock some cells such that disruption occurs. To enhance the binding together of the particulates the pH is adjusted down towards the isoelectric point of the proteins, that being typically in the range 3.5 to 5.0 pH units.

Twelve percent w/w solids containing float blankets can be achieved in a well-designed and well-operated flotation stage. The effect this has on water removal requirements may be seen after examining the following expression:

$$\begin{matrix}\text{Amount of water} \\ \text{to be removed after} \\ \text{flotation stage}\end{matrix} = 24.6\left(\frac{1}{0.12} - 1\right) - \left(\frac{1}{0.92} - 1\right)$$

$$\begin{aligned} &= 24.6 \times (7.33 - 0.087) \\ &= 24.6 \times 7.24 \\ &= 178 \text{ t/hr} \\ &= 7.24 \text{ t/t cells} \end{aligned}$$

Now the cost of drying at £3.5/t water evaporated is £25/t cells.

A 12% w/w paste of cells and cellular debris is not at the limit of handleable viscosities, and since viscosity is the only constraint on further mechanical dewatering, another stage may be added to the process to remove more water prior to thermal evaporation. A centrifuge will concenrate the 12% w/w paste to 20% w/w with the rejected water being available for recycling to the fermentor stage just as would be the case with underflow culture from the flotation stage. There is a solids concentration for flocculated materials above which it is not possible to get by purely mechanical dewatering means alone. The reason for this is that a matrix of linked flocs forms which very effectively resists enhanced gravity forces (in the case of centrifuges), thus preventing liquid from being removed from the interstices. Almost all industrial dewatering operations meet the same limit for differing reasons, at which point thermal drying becomes the only means of removing more moisture. For the above reason and also because of the difficulty of pumping sludges of greater than 20% w/w solids which have highly non-Newtonian Rheology, 20% is the practical limit of mechanical dewatering.

The beneficial effect of including such a stage in the process is clear from considering the calculation below:

$$\text{Amount of water to be removed} = 24.6\left(\frac{1}{0.2} - 1\right) - \left(\frac{1}{0.92} - 1\right)$$

$$= 24.6 \times 3.91$$
$$= 96.3 \text{ t/hr}$$
$$= 3.91 \text{ t/t cells}$$

So the final cost of drying = £13.7/t cells

What will be the chief raw materials requirement? At this stage, the fermentor size and flow rate have been roughly determined and the mass balance around flotation and centrifugation has been calculated approximately. The drying section evaporative load is also known in round figures.

As has already been suggested, both the underflow from the flotation stage and also the centrate from the centrifugation stage can be returned to the fermentor to make up the volume of liquid removed in harvesting the cells. A detailed mass balance would come later. By returning most or all of the underflow or centrate, it is possible to keep trace nutrient losses down to the minimum. This reduces the required input rate to maintain healthy cell growth of each nutrient compared with the case where nutrients are rejected as effluent in, say, a rejected centrate stream.

Individual mass balances on trace nutrients can be carried out if the cellular mass fraction of each element or compound is known. For instance, if the PO_4^{3-} level in cell is 0.077 (kg/kg) then the PO_4^{3-} level in product is calculable by adding the cellular PO_4^{3-} quantity to the soluble PO_4^{3-} quantity in the water associated with the feed paste to the driers.

Say, mass fraction of typical drier feed cream:

0.19 cell particles (insoluble)
0.01 soluble organics & inorganics
0.80 water
1.00

If, say, the soluble organics and inorganics fraction contains 0.15 (kg/kg) PO_4^{3-}:

$$= 0.19 \times 0.077 + (0.01 \times 0.15)$$
$$= 0.016 \text{ kg/kg of feed cream}$$

$$= 0.8 - \left(\frac{0.2}{0.92} - 0.2\right)$$
$$= 0.78 \text{ kg/kg feed cream}$$

$$= 1 - 0.78$$
$$= 0.22 \text{ kg/kg feed cream}$$

$$= \frac{0.016}{0.22}$$
$$= 0.073 \text{ kg/kg}$$
$$= 7.3 \text{ w/w}$$

Now, having calculated the exit product PO_4^{3-} level it is a simple job to work out the input phosphate required as, say, H_3PO_4 to satisfy product needs. This mass balance only works provided there is no other loss of PO_4^{3-} from the process as there would be if, say some of the centrate stream was diverted to effluent. In addition, the soluble organics and inorganics mass fraction of PO_4^{3-} can only be determined once the metabolic requirements of the cell are quantified. This quantification will allow concentration levels of nutrients to be determined in the fermentor. For the initial rough mass balance, therefore, it is probably appropriate to omit the solubles contribution.

The major requirement of the cells in terms of raw materials is for methanol. Laboratory data must be consulted in order to understand the likely value of the ratio:

$$\frac{\text{methanol consumption}}{\text{cells produced}}$$

It must be remembered when studying laboratory results that the ratio will most likely be expressed in terms of washed and dried whole cells produced from a laboratory fermentor. When translating the figure to a form applicable to the large scale, account must be taken of all the soluble organics and inorganics which are dried onto the cells. In other words, on the large scale, the ratio is expressed as:

$$\frac{\text{t methanol input to plant}}{\text{t product produced}}$$

and this ratio will always be lower than the laboratory figure which was expressed in terms of washed and dried whole cells. For *Methylophilus methylotrophus*, large scale plant methanol/product ratios are in the range 1.8 to 2.1.

Having now dealt with water, phosphate, and methanol, the other main raw material is oxygen, all other materials such as ammonia, magnesium, and sulfur being used at considerably lower rates than water, phosphate, methanol, and oxygen. To understand the extent of the organisms' need for oxygen and also to help check the consistency of laboratory data such as methanol utilization, it is well worth spending time developing a simple chemical equation for metabolism of cells. Typical microorganisms analysis can be simply expressed as follows:

carbon content mass fraction	0.470
nitrogen content	0.140
phosphorus content	0.020
oxygen content	0.300
hydrogen content	0.070
	1.000

Very crudely, therefore, the chemical formula is $CH_2O_{1/2}N_{1/3}P_{1/60}$. Since the P contribution is so small, this can be ignored. So the formula becomes:

$$CH_2O_{1/2}N_{1/3}$$

Now, using the laboratory data that states 1 unit of methanol for, say 0.5 units of cells, the equation balanced for carbon containing species only can be written:

$$2CH_3OH \rightarrow 1.19CH_2O_{1/2}N_{1/3} + 0.81\ CO_2$$

Mwt	32	27
Mass	64	32

Another very frequently used concept is that of carbon conversion efficiency. This is a percentage expression of input carbon actually incorporated into the cellular structure. In the case above, carbon conversion = 59.5%. For reasons of simplicity, the carbon conversion will be assumed to be 50%, then the full chemical reaction may be written starting with the usual reactants and finishing with the usual products:

$$0.3\ NH_3 + 2O_2 + 2CH_3OH \rightarrow CH_2O_{1/2}N_{1/3} + CO_2 + 3.5H_2O$$

Observe now that while the carbon conversion is 50% to cells, the methanol to cells ratio equals:

$$= \frac{64}{27} = 2.37$$

The requirement for aeration can now be calculated. Oxygen requirement per te cells = $\frac{64}{27}$ = 2.37 te/te cells, but the O_2 content, in gravimetric units, of air = 0.23 so air requirement per te cells = $\frac{2.37}{0.23}$ = 10.3 te air/te cells.

but cellular production	= 24.6 te/hr
. . air requirement (to meet O_2 need)	= 246 × 10.3
	= 253 te/hr
and expressing in a standard volumetric way	$= \frac{253}{29} \times 22.41$
	= 196,000NM³/hr

In fact the total air supply to the fermentor needs to be substantially greater than 196,000 NM^3/hr since the O_2 supplied will be utilized only to a certain degree. The concept of O_2 utilization efficiency will be explained later.

What does the process look like overall? At this point, and without delving any deeper into the details of each processing stage, an overall view of the scale of the process is available. The salient features brought out so far can be shown on a sketch of a flowsheet of the process.

200,000 te pa SCP from methanol plant:

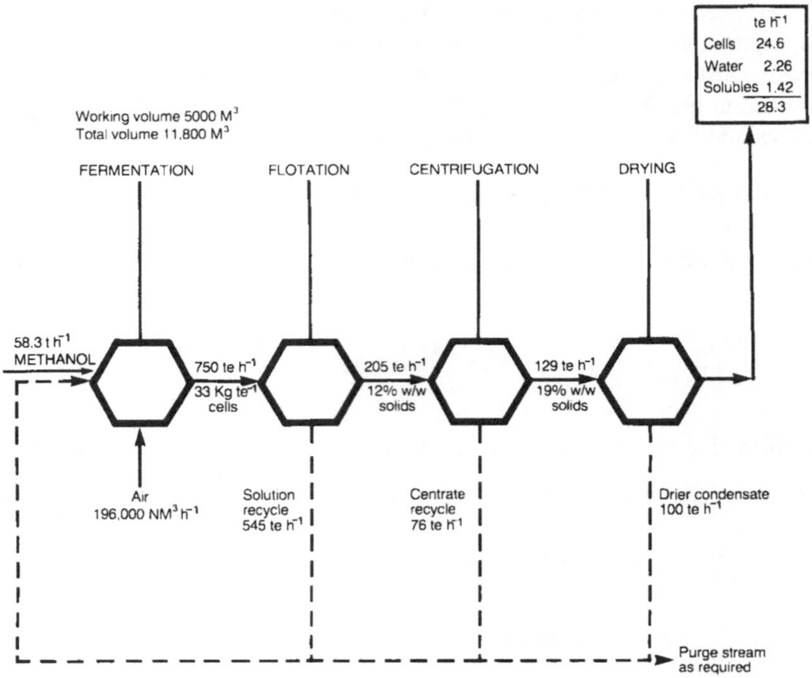

No self-respecting engineer would be satisfied with the accuracy or consistency of the flowsheet shown. Further work would be needed to balance all input species with output species, taking into account the organism metabolic equation. However, at least the scale of the elements in the process can be sensed.

If desired, it should be possible at this stage to do a rough estimate of the variable cost of production in that the substrate usage is known, the drier evaporative load is known, and a stab could be made at assessing power requirements in drying and centrifugation. The power requirement for aeration is more difficult to assess until the total air rate and pressure are specified. Now, returning to the concept of oxygen efficiency the engineering objective in designing the fermentation stage must be to minimize both the amount of air required to be compressed as well as the air pressure. Later on in the section on fermentation the subject will be tackled more fully but for now an assumption will be made that a 30% oxygen efficiency can be achieved. In other words, 30% of input oxygen is transferred usefully to the culture and cells within the fermentor.

$$\text{Now the total air requirement is } 196{,}000/0.3 = 653{,}000\text{NM}^3/\text{hr}$$

It will be demonstrated later that to reach such an oxygen efficiency a high air pressure must be used. Assuming a 9 bar/g delivery pressure, the adiabatic minimum power required (in a four stage machine) is given by the following formula:

$$= nP_1V_1\left[\frac{\gamma}{\gamma - 1}\left(\frac{P_2}{P_1}\right)^{\frac{n-1}{n}} - 1\right]$$

$$P_1 = \text{inlet pressure} \quad N/m^2$$
$$V_1 = \text{inlet volume} \quad M^3sec^{-1}$$
$$P_2 = \text{delivery pressure} \quad N/m^2$$
$$n = \text{no. of stages}$$
$$\gamma = 1.4 = \frac{Cp}{Cv}$$

$$= 4 \times \frac{653{,}000}{3{,}600} \times 10^5 \times \frac{1.4}{0.4}\left(10\left(\frac{0.4}{5.6}\right) - 1\right)$$
$$= 2.539 \times 10^8 (10^{0.071} - 1)$$
$$= 2.539 \times 10^8 \times (1.178 - 1)$$
$$= 0.45 \times 10^8 \text{ J/sec}$$
$$= 0.45 \times 10^8 \text{ watts}$$
$$= \underline{45 \text{ MW}}$$

Allowing, say, a 40% loss due to machine inefficiency the power need is 75 MW.

Finally, in order to demonstrate the cost of aeration, if electricity is used to drive the compressor at, say, £35/MWH then the electricity cost per tonne of product

$$= \frac{75 \times 35}{28.3}$$
$$= £93/t$$

C. The Processing Stages

1. Media Preparation

In this section the supply of nutrients necessary to support healthy organism growth is considered. The first question that arises is: what is the nutrient recipe necessary to support cell growth? R&D data needs to be consulted. A continuous culture recipe will include all essential elements such as P, S, Mg, Ca, Fe, K, Cu, Mn, etc., in the form of salts or as acids. These salts or acids are then proportioned to a carbon substrate requirement to give the liquid nutrients ratio to carbon source. Now, when a need exists for reducing the pH of the harvesting stream to assist in the preparation of cells for flotation, an opportunity presents itself. By adding, say, S or P as H_2SO_4 or H_3PO_4, respectively, not only is the element supplied to support metabolism but the acid also produces the required pH drop. Phosphate is returned to the fermentor after the flotation stage in the underflow and in the centrate.

Since the metabolic processes in the cell tend to cause a drop in the culture pH and also because acidic underflow and centrate are being returned to the fermentor, pH adjustment needs to be carried out.

By adding N as NH_3 gas in the air stream the facility to adjust the culture pH is provided. Clearly other possibilities exist such as adjusting pH by KOH and supplying nitrogen from $(NH_4)_2SO_4$. As long as the recipe requirements are met and nutrient process effects such as pH raising and lowering are achieved, there is some flexibility available to the engineer in fixing the formulae of nutrients. However there are also constraints such as that imposed by the need to avoid chloride ions in the presence of most stainless steels.

Three further examples of constraints are (1) product analysis specification; (2) any undesirable salt complexing at certain ionic concentrations; and (3) solution buffering effects which may cause problems when attempting to raise and lower pH. In the latter case there is a well-defined problem area when using phosphate ions in solutions.

Having established the form and ratio of nutrients required the information needs to be translated into culture standing concentrations in ppm or g/ℓ of each element or radical.

Then the tolerance of the organism both to reduced and increased concentrations must be established and documented. For instance, it is known that above 1500 ppm of NH_3 an inhibitory effect can be produced on metabolic rate to biosynthesis. Similarly very high SO_4^{2-} concentration causes distress to the organism. Obviously any single nutrient can be made to limit metabolism at low enough concentrations and since a C-source limited fermentation is required, by definition oxygen and all other nutrients must be in excess.

The engineer is heavily dependent upon data produced in the laboratory at this juncture. A two-way communication has to take place to ensure that the right laboratory analyses are performed in order that the total plant nutrient mass balance can be constructed. Finally, a qualifying comment on the above discussion can be made. When considering nutrients, it is the inorganic salts and acid requirement, together with any organic growth promoters, that are being referred to. Methanol supply is a subject in its own right and will be dealt with in Section III.C.3. Oxygen supply, likewise, has peculiar problems associated with it and is also dealt with in Section III.C.3. A requirement for antifoams (which undoubtedly there will be) is going to be apparent early on in fermentation experiments in the laboratory. The standing concentration must be defined such that a mass balance can be constructed.

A full nutrients mass balance should be calculated now, not only for the design condition but also for the range of operating conditions likely to be encountered in plant running. One or two elements will need to be included which have not yet been mentioned, such as titration data for culture in order to predict the acid addition rate required to effect a given pH reduction.

How are the nutrients to be added to the fermentation stage? Salt solutions or acids will be stored in tanks on site and pumped at the required rate through some sterilization system into the fermentor.

The point of addition of the nutrients matters little but it is preferable in order to reduce the chance of heavy complex salt precipitations around nozzles, to introduce the nutrients in the acidic recycled water stream which will also contain recycled nutrients and phosphate. On the other hand, direct addition to the fermentor may be preferred in that during plant start-up in batch growth conditions it might be thought desirable to have direct addition of nutrients. Storage tanks need to be sized, stirred, and heated as appropriate, depending upon the stored material.

Are there any special features associated with start-up? In the unlimited logarithmic growth phase when an initial inoculum of cells is being grown up to allow the cell concentration to reach the design (32.8 g/ℓ in this case) it is possible that the cells will require a slightly different nutrients ratio than for steady state operation. This is due in part at least to the change in cellular composition with growth rate mentioned earlier on in Section III.B.

2. *Sterilization*

There are two methods of sterilization available: filtration and thermal. Other processes exist such as UV irradiation but as yet no new process offers more benefits than those stated above.

Which means of sterilization is applicable to which streams? The simplest and cheapest (in variable cost terms) method of sterilization is filtration. In every possible case, therefore, filtration should be applied. However, cases outside the realms of possibility for filtration include all streams which are not clean liquids or gases. Unfortunately apart from refined methanol, air, and ammonia, all other input streams are dirty in respect to handleability in sterile filters. Thus all salt solutions, acids, and antifoams are heat sterilized.

The detail relating to design of wet heat sterilization methods is well documented and will not be considered further except to say that in principle a wet heating section must be followed by a wet retention section in order to give an adequate contaminant kill. The contaminant kill is measured in terms of the reduction in number of viable cells. Wet conditions must prevail throughout.

In filtration, again, apart from the unusually small pore size (down to <0.2 μm in some cases) there is nothing different to conventional filtration. Therefore further consideration of it will not be given here.

Next in this section a brief mention needs to be made of the engineering aspects of sterilization. While it is possible to weld up most equipment, thus eliminating flanges and possible areas of contaminant ingress, it will always be necessary to engineer sterile standard joints for such things as vessel manways and pressure pads. The main aim is to create very smooth surfaces between which a suitable jointing material made from say, synthetic rubber, can be sandwiched. It is only by ensuring very high standards of finish that such a joint can then be termed sterile.

Finally, consideration should be given the saturated steam requirement to sterilize the total fermentation and ancilliaries system prior to fill up and inoculation. Wet steam must contact and raise to temperature every surface of the equipment and prior to this all air and inerts must be displaced. The quantity of steam required is not inconsiderable for large-scale fermentations.

3. Fermentation

The fermentation conditions so far defined are

1. Methanol rate required to give the output
2. Approximate air rate requirement based on an assumed oxygen utilization efficiency
3. Size of vessel
4. Concentrations and rates of all nutrients

The objective in fermentation is to supply all the nutrients required by the cell in order to let it grow healthily, at the same time removing satisfactorily any metabolic by-products such that growth inhibition does not occur. In the latter category the following must be considered: any gaseous products (mainly CO_2), any soluble products excreted, and the heat produced by biosynthesis.

It turns out that the fermentation needs can be reduced to three prime requirements, namely oxygen transfer, mixing, and heat removal. Dealing with each requirement in turn, the first question to be asked relates to supply of oxygen to the fermentation, namely: how much air must be supplied to the fermentor in order to satisfy oxygen metabolic needs, given a limited ability to transfer oxygen into the liquid? The phrasing of the question is highly significant: its formulation recognizes that the process of supply of O_2 to fermentation cultures is mass-transfer-controlled at the gas/liquid interface. To provide an answer the concept of oxygen efficiency must be introduced.

Oxygen efficiency is defined as

$$\frac{\text{total oxygen used by organism}}{\text{total oxtgen supplied to fermentor}}$$

It can be measured for a given fermentation by recording the air inlet rate together with gas exit rate and O_2 analysis. The problem the designer has is to determine the oxygen efficiency for a fermentor which is yet to be constructed.

Once the oxygen efficiency has been determined, it is a simple matter to calculate the required total air rate since the metabolic requirement of the culture for oxygen will already have been measured on the laboratory scale.

Typical laboratory derived data for O_2 metabolic requirement is as follows: oxygen metabolic requirement = 5.2 (m moles O_2/l min) where 1 refers to ungassed culture.

A consistency check may be carried out, incidentally, on the laboratory produced O_2

metabolic requirement by comparing this with that given in the stoichiometric equation developed earlier.

For example, laboratory data says O_2 metabolic requirement = 5.2 m moles $O_2/\ell/min$. Using the equation derived in Section III.B., oxygen requirement = 2.37 t/t cells; = 2.37g/g cells; but cell concentration = 32.8 g/ℓ. Therefore, oxygen requirement = 32.8 × 2.37 g O_2/ℓ, but dilution rate = 0.15 hr^{-1}. Oxygen requirement = 11.66 g $O_2/\ell/hr$; = 11.66/60 × 1000/32 m moles $O_2/\ell/min$; = 6.07 m moles $O_2/\ell/min$. Compare laboratory data which says O_2 metabolic requirement = 5.2 m moles/ℓ/min.

The unresolved problem is to determine the oxygen efficiency. Oxygen efficiency is dependent upon two variables, (1) rate of transfer of O_2 into the liquid phase, and (2) length of time that the transfer is allowed to take place.

Classically the rate of transfer of O_2 into the liquid phase is defined as being proportional to the driving force for transfer of O_2 and to the inverse of the resistance to transfer. Rate of transfer of O_2 = $K_L a\overline{\Delta}C$.

Three questions can now be asked and by answering these questions it will be possible to determine oxygen efficiency for the fermentor being designed. The questions are

1. How is the driving force for oxygen transfer determined?
2. How is the resistance to oxygen transfer determined?
3. How is the residence time for O_2 transfer established?

The driving force for oxygen transfer is simply the difference between the average equilibrium O_2 concentration in liquid under the conditions that apply and the actual (as opposed to equilibrium) O_2 concentration within the liquid under the same conditions. The average equilibrium O_2 concentration at the liquid surface is proportional to the O_2 partial pressure within the fermentor and hence it can readily be determined for a given fermenter by calculating the average (log mean) O_2 partial pressure in the gas phase. Clearly, both the depth of culture in the fermentor and the pressure above the culture will influence the average O_2 equilibrium partial pressure. Generally, higher driving forces for O_2 transfer can be achieved in tall vessel operating under pressure.

A more knotty problem is to decide upon the actual liquid O_2 concentration that will be achieved in the fermentor. The limit is reached in driving force potential when the actual liquid O_2 concentration falls to zero. However, methylotrophs cannot tolerate a zero o_2 concentration and just as for other nutrients a certain minimum O_2 concentration (normally called C_{oxygen} critical) is essential. This minimum concentration must be determined in the laboratory and it can be used to estimate the likely driving force. However, the value for the actual liquid O_2 concentration can only be determined by carrying out an O_2 transfer mass balance when the known organism O_2 requirement (say 5.2 mmol O_2/1 min) is equated with the O_2 transfer rate. In addition the value of the actual liquid O_2 concentration is not calculable from the equation unless the resistance term is quantified.

The resistance to O_2 transfer is defined in terms of an overall average mass transfer factor per unit volume of fermentor. Conventionally this is expressed using the following symbols:

$$= k_L \times a$$

$$= \text{mass transfer coeff.} \times \frac{\text{interface area}}{\text{unit volume}}$$

k_L units: (m/sec)
a units: (m^2/m^3)

Determining the value for $k_L \times a$ is a lengthy subject in its own right and is dealt with in many textbooks. However, the approach to defining $k_L \times a$ for a particular fermentation is briefly as follows. k_La is proportional to the power input per unit volume of fermentor, superficial air velocity through the fermentor, culture temperature, and finally surfactant concentrations and culture properties. So, for each fermentation, k_La must be determined on the small scale. Then making the assumption that culture properties, temperature, and concentrations of surfactants will be invariant with scale, it is possible to scale-up the fermentation stage to the size required by the designer using the following relationship:

$$k_L \times a \text{ proportional to } \frac{p^\gamma}{V} \bullet (V_s)^\delta$$

$$V_s = \text{empty fermentor air velocity}$$

where the value of k_La already determined in the laboratory is held constant for any scale providing the p/V and V_s terms are adjusted accordingly. (Note that for large air lift fermentors where no mechanical power is dissipated, a relationship has to be assumed between mechanical and aeration power. Such relationships are available with stated assumptions in the textbooks.)

The residence time for O_2 transfer is determined by the contact time between gaseous O_2 and the liquid. Contact time can be maximized by ensuring small bubble size distribution and having as tall an air lift column as possible. However, it is undesirable to take the concept too far. If the bubbles are very small and the column is very tall then all the available O_2 within a bubble will be utilized before the bubble reaches the liquid surface. Hence the bubble will effectively provide no O_2 for the last part of its rise through the liquid column. Contact times for bubbles can be calculated if a mean bubble diameter is assumed and if something is known about the bubble slip velocity through the fluid. Measurements made in the culture in the laboratory will provide invaluable information on this subject but it is essential that the laboratory uses the culture that will actually exist in the large-scale vessel for test purposes.

The three preceding sections have shown how each element in the equation of O_2 efficiency is determined. It will have become clear upon reading through the above sections that a knowledge of the height and configuration of the fermentor is implied. For example, the contact time to determine O_2 residence time is dependent upon the liquid column height. The designer knows the overall working volume he requires for the fermentation and therefore he can dimension the vessel as he pleases. However two further areas need to be mentioned. First, the assumption about gas holdup must be validated for the fermenter dimensions chosen. In Section III.B it was stated that a gas holdup between 0 to 45% may need to be catered for. In the literature various relationships are found relating gas holdup Σ to superficial gas velocity but a more fundamental and satisfactory approach can be taken. If the bubble size distribution can be predicted, then it will be possible to calculate an average effective bubble slip velocity through the culture (making due allowance for any culture velocity in the first place). Then by deriving an equation as below:

$$\sum \text{ proportional to } \frac{\text{Air rate to fermentor (m}^3\text{/sec)}}{\text{cross sectional area of fermentor (m}^2\text{)} \times \text{average bubble rise velocity (m/s)}}$$

the likely value of $\sum$ may be determined.

Second, and linking with the discussion on bubble slip velocity, there is a need to consider the efficiency of the disengager device used to remove the spent gases from the culture. No disengager will be 100% efficient at removing all of the bubbles approaching it. The smallest bubbles (and by definition the most difficult to disengage) will slip through the disengager only to be swept back down into the culture. A detailed study of the culture properties is necessary to determine the degree of difficulty of removing bubbles from within the liquid bulk. Only then will it be possible to predict disengager performance and determine volume of undisengaged gas bubbles.

Mixing within the fermentor — what are the requirements? All fermentations benefit from thorough mixing, mainly because mixing ensures a totally uniform nutrients concentration throughout the vessel. Nutrients can be organic and inorganic but in addition dissolved O_2 can equally well be considered as a nutrient to be mixed throughout the vessel. As the earlier discussion on O_2 transfer suggests, mixing is essential in order to rapidly renew surfaces of bubbles and so avoid equilibrium being established in the liquid close to the gas/liquid interface.

As well as transferring nutrients to the organism, mixing is also equally essential in order to transfer metabolic products and waste gases (such as CO_2) away from the organism and into the bulk culture. Any local inhomogeneity results in a reduction of fermentation efficiency. So if mixing is so essential, how is the minimum desirable level to be determined so that a balance is struck between the benefit of more mixing against the additional costs?

The simplest way to quantify mixing is by determining the power input per unit volume of the fermentation (i.e., the power input per unit gassed volume). Scale-up can be done by maintaining constant power input per unit volume although in reality it is likely that less energy can be afforded to be expended on a large scale per unit volume than is the case in a small laboratory unit.

In the case of the air lift fermentor, all power input is derived from the expansion of gas through the column of culture. This expansion causes circulation of the culture and the energy is expended as frictional pressure losses in the vessel.

If, say, 1 kW/m^3 gassed volume is to be expended in order to achieve confidence in the ability to mix sufficiently, then in the fermentor under consideration, 9.09 MW of power is effectively being used. A cross check should be done on the expected power input from the calculated total air rate to see if indeed 9.09 MW will be expended by expanding the air.

Mixing can always be made more effective if an attempt is made to distribute nutrients evenly throughout the fermentor in the first place. This can be done for instance by creating a few jet mixing points or by some sort of manifold system within the vessel.

Heat removal from the fermentor — how is it to be achieved? Heat is removed by passing culture over cooling surfaces. The surfaces are normally those of tubes arranged in bundles through which a coolant (normally just cooling water) flows.

The design requirement is to size the tube bundles in terms of the number required and also the available surface area of each bundle.

The calculation is a straightforward chemical engineering exercise utilizing the well-known formula:

$$Q = U_oA \cdot T_{LM}$$

The only difficulty with using the expression is met when trying to determine the value of U_o (kW/m^3K) — the overall heat transfer coefficient.

Biological fouling is a well-known phenomenon in process heat transfer and is it the major cause of poor heat transfer performance in fermentors. The air lift fermentor employs a configuration which ensures a very high process fluid (culture) velocity passed the heat transfer bundles. The high velocity is a benefit since the value of U_o can be shown to be proportional to velocity: U_o proportional to $Re^{0.8}$ where Re = Reynolds number. More importantly, the high velocity causes shear on the tube walls and this very effectively scours any buildup, thus keeping the tubes clean. The so-called dirt film factor in the value of U_o is thus kept low.

Tests should be carried out on real culture passing over a heat transfer surface for prolonged periods of time to assess the fouling that takes place. Then by considering the other elements that go to make up the composite U_o the average achievable value may be determined.

4. Harvesting

Harvesting of cells is carried out first by exploiting the property of the culture having subjected it both to acid and heat shock, and second, by mechanical dewatering in centrifuges. A clear logic exists in carrying out the processes in the specified order, since in going from 3% w/w to 12% w/w in a flocculation/flotation stage the amount of water to be removed is

$$\text{water removed in flotation} = \left(\frac{1}{0.03} - 1\right) - \left(\frac{1}{0.12} - 1\right)$$

$$= 32.3 - 7.33$$

$$= 25 \text{ t water/t cells}$$

whereas in the centrifugation stage, going from 12 w/w to 20 w/w the moisture to be removed is

$$\text{water removed in centrifugation} = \left(\frac{1}{0.12} - 1\right) - \left(\frac{1}{0.2} - 1\right)$$

$$= 7.33 - 4$$

$$= 3.33 \text{ t water/t cells}$$

Now, in flotation, no energy is expended except for the thermal energy of heat shocking (which by careful engineering can be largely recovered in a heat interchanger) and the chemical energy of acid shocking (resulting in a small consumption of acid). However, in centrifugation about 1.7 kW/m^3 hydraulic feed is expended in a centrifuge of the disc stack type and this figure assumes a flocculated feed material to start with. If the particle size is lower than a flocculated feed material then the energy consumption will be even greater.

Design of the flocculation stage will be based very much on experience gained from semitechnical trials. It is impossible to predict performance and size vessels without carrying out a test program to simulate the process. Aspects to be considered, however are listed below:

1. Amount of gas released by changing the pH of the culture
2. Amount of gas released by depressuring fermentor culture (assuming that the flotation is carried out at atmospheric conditions) — it is the gas being released that causes the process of flotation to occur, and an understanding of how the gas bubbles nucleate and grow is important to the designer
3. Amount of acid needing to be added to produce the required pH change

4. Amount of steam needing to be added to produce the heat shock effect
5. Rate of rise of floc blanket. (This is needed to allow the flotation tank cross-sectional area to be sized)

Conceptually, designing a flotation stage is almost identical to designing a disengager for gas liquid systems. No detail will be entered into on the centrifugation stage since centrifuge manufacturers supply all the necessary test facilities and scale-up criteria for the engineer to enable a process design to be carried out. However, it must be noted that special attention should be paid to the risk of floc damage caused by high shear rates in such things as tank agitators, pumps, and centrifuges themselves.

5. Drying

A 20% w/w feed cream has been proposed as the feed material to the drying section. The immediate question to be addressed is: what type of drier is required? A supplementary question then arises, namely: what form of product is desired? If a powder product is required, then the feed cream can be directly spray dried with the appropriate degree of atomization in order to produce a powder in the correct particle size range. However, alternative forms of drying could equally well be used, for instance, rolling bed granulator driers, fluid bed driers, or possibly scraped surface evaporators, and the product from the chosen drier could be ground in mills to the correct particle size. Equally, it may be more desirable for marketing reasons to produce a granular or pelleted form of product. In these cases, driers mentioned above would be suitable without the inclusion of a grinding stage, or some other form of drier may be found which gives a uniform large particle size in the finished product.

The ultimate decisions on the type of drier will depend upon the eventual form of product desired. However, the energy requirements of each drier type must be known so that when the time for selection arrives, due regard can be paid to eventual likely energy costs. Generally speaking, the higher the differential heating medium temperature across the drier system (inlet-exit gas temperature for a gas-fired hot air drier, for instance) the lower is the heating cost expressed per ton of finished product.

This effect is a direct and unavoidable consequence of thermodynamics. Consider two drier systems running on natural gas and consider the difference in gas usage per ton of water evaporated (the figures are approximate):

Inlet Temp	**Exit Temp**	**Gas usage/te water evaporated**
500	120	80 m^3/te
300	120	100 m^3/te

The conclusion could be drawn that high inlet drying temperatures are always desirable to minimize costs. Unfortunately there is a further element to be taken into account. High protein content materials suffer from heat degradation and the amino acid profile can be adversely affected by prolonged or excessive heating. Thus with respect to product quality, lower temperature treatment is more desirable. Clearly the choice of drier depends upon a few variables:

1. Desired product form and the ability of the drier under consideration to produce something similar
2. Cost of running the drier (both from the heating and the power point of view)
3. Degree of thermal degradation of product that can be tolerated before the product quality suffers

As for centrifuges, manufacturers of drying systems will quote energy costs and are normally prepared to carry out tests for clients considering their driers. It is only by weighing the merits of each potential suppliers' equipment that a fit will be obtained between the required duty and the best equipment on the market to carry it out.

6. Product Handling

Both powder product and granular products can be handled in conventional equipment, for instance, rotary valve conveyors, pneumatic conveying systems. Conveyor belts probably are inappropriate because of dust emission problems. Dust control must be given detailed attention for two reasons. Using the well-known three-cornered diagram:

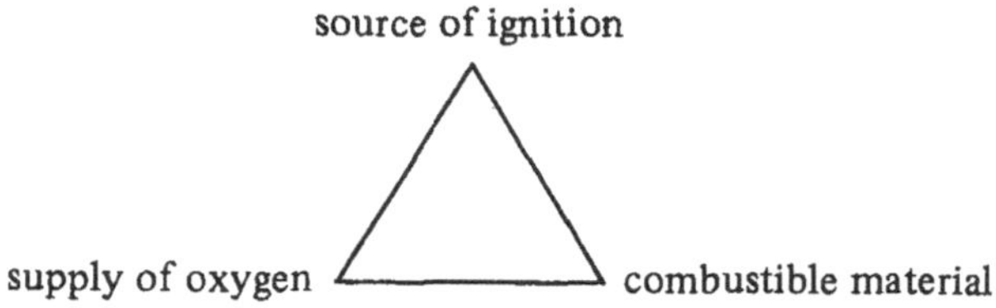

In all cases when handling biomass, a combustible material is present. Unless inert atmosphere systems are used, then oxygen is also present. In order, therefore, to prevent an explosion or rapid thermal decomposition, it is essential that no source of ignition can occur. Sources of ignition can be sparks generated electrostatically or by mechanical means or hot or smouldering contaminants dropped into the product.

There is a way open to the designer, however, whereby he can effectively render the product incombustible. This can be done by defining the dust concentration levels above and below which it is *not* possible to cause a fire or explosion by adding a source of ignition in the presence of oxygen. The lower concentration limit may not be able to be met and therefore may not be of much use in practical terms. However, dense phase conveying, as it is termed, is a fairly well established means of avoiding the combustion concentration region.

What happens in drying where there is a combustible product and a ready source of ignition (by definition)? Drier manufacturers handle the problems created in drying potentially explosive products in many ways. Perhaps the best, however, is the recirculation of spent combustion gases which will be rich in CO_2 and N_2 but will have as little as 5 v/v O_2. These spent gases are used as the heat carrier back to the drier. In other words, gas and air are combusted and the heat is transferred into the recycled gas stream which is then used to contact with the product. In this way the supply of O_2 is limited to acceptable levels.

D. Other Companies' Choice of Reactor

Mitsubishi Gas Chemical, Hoechst, and Norprotein all were persuaded to follow a similar choice of reactor.[1,3,12] The only large-scale example of stirred tank reactor for single-cell protein was the BP process on *n*-alkanes. Figures 2, 3, 4, and 5 show the flowsheets for the various processes suggested for SCP production from methanol. Figure 6 shows the ICI Plant at Billingham.

E. Toxicological and Nutrition

Toxicological and nutritional trials are very important in the development of novel foodstuffs. From the earliest use of *Methylophilus methylotrophus* experimental studies have been carried out both within and outside ICI to establish its suitability as a nutritious food supplement. Classical toxicological studies included multigeneration trials using laboratory animals and target species (e.g., fowl). No evidence was found of pathogenicity, toxicity,

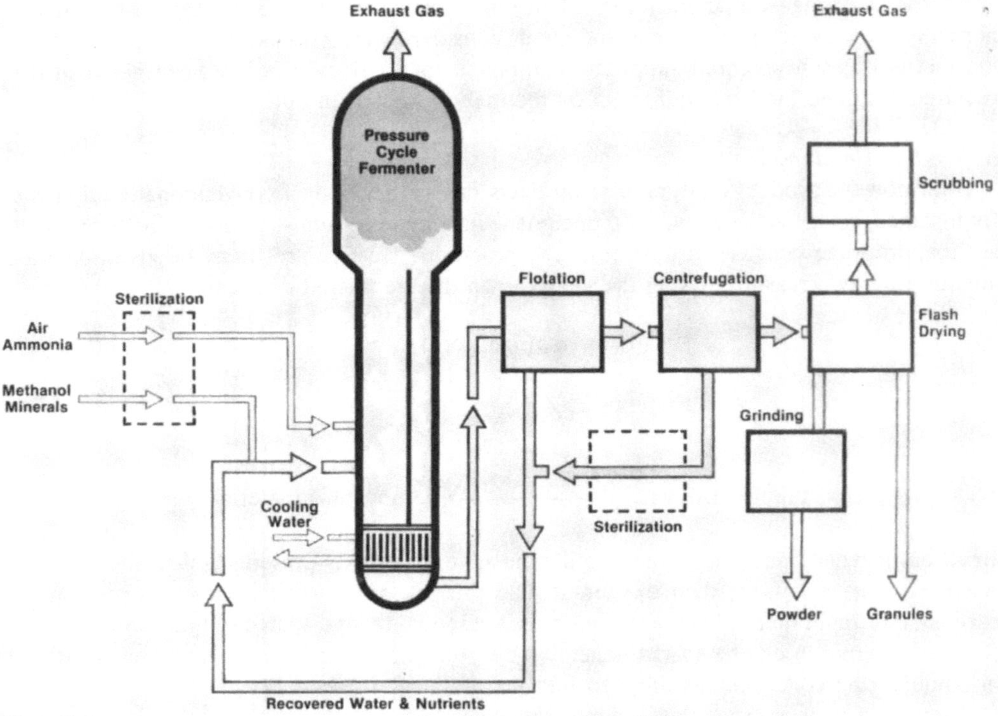

FIGURE 2. Flowsheet of ICI plant at Billingham.

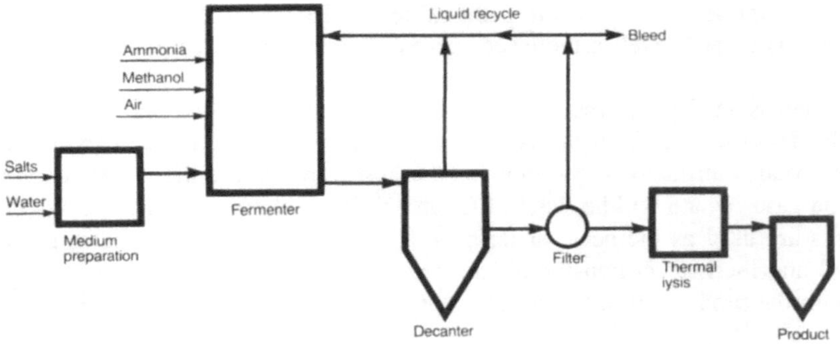

FIGURE 3. Suggested flowsheet process of L'Institute Du Petrole.

mutagenicity, or teratogenicity. Over 250,000 animals have been fed on Pruteen (Table 1) including fowl, veal, pigs, fish, and mink. Extensive evidence has been accumulated to show that Pruteen is a valuable foodstuff. We refer the reader to the paper by Stringer[13] for further details.

Table 2 compares the crude analysis of the ICI product with pilot scale product from other processes. Table 3 compares the amino acid content of each SCP as published.

IV. CONCLUSION

In 12 years from 1969, SCP on methanol was developed from the shake flask to full-

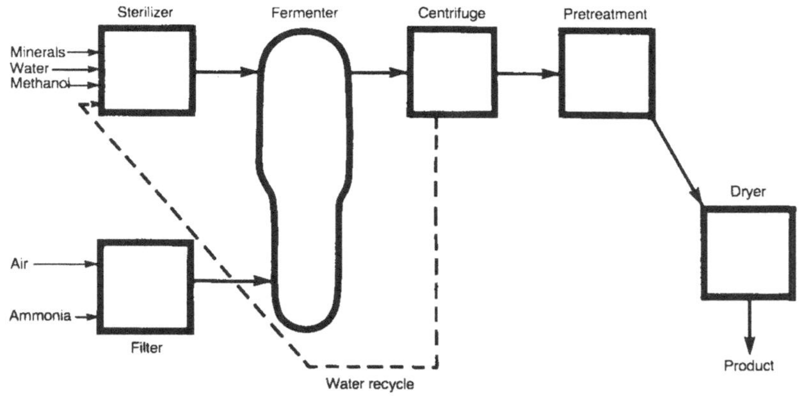

FIGURE 4. Suggested flowsheet process of Mitsubishi Gas Chemical.

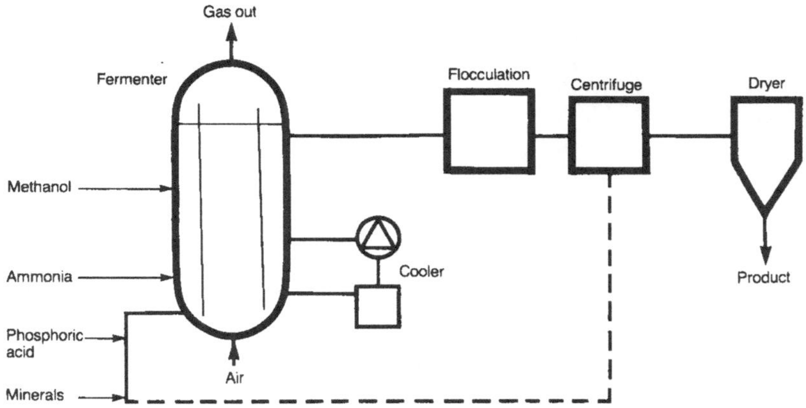

FIGURE 5. Suggested flowsheet of Hoechst-Uhde.

scale fermentor operation. ICI now has the world's largest fermentor in operation, operating continuously for periods in excess of 100 days. The future for SCP now depends on economic factors rather than microbial factors.

This technology may be developed in parts of the world where gas is available and where high quality protein is seen to be required.

There is no doubt that the development of this process has crossed into new realms of technology in the large-scale cultivation of microorganisms.

FIGURE 6. The ICI PRUTEEN plant.

Table 1
ANIMAL PRODUCTION TRIALS — PRUTEEN

Species	Trials	Animals
Broilers	49	173,300
Layers	6	10,530
Turkeys	10	10,720
Piglets	19	3,750
Pigs	14	1,032
Fish	16	109,100
Veal calves	41	3,750
Calves	22	1,250

Table 2
AMINO ACID CONTENT OF VARIOUS SCPs

	g/16 g N[a] PRUTEEN	g/16 g N[b] MPC (yeast)	g/16 g N[b] MPC (Bact)	g/100 g product NP	g/100 g[d] HP
Alanine	7.72	5.91	6.9		
Valine	5.89	5.19	5.4	4.7	4.3
Glycine	5.45	4.32	5.5		4.3
Isoleucine	4.99	4.96	4.5	3.7	3.6
Leucine	7.61	7.19	7.1	5.9	6.7
Proline	3.33	3.57	3.5		
Threonine	4.77	4.50	4.3	3.6	3.6
Serine	3.55	4.56	3.4		
Methionine	2.34	1.58	2.3	1.7	1.9
Phenylalanine	3.94	4.45	3.8	2.9	3.8
Aspartic acid	9.06	8.84	8.9		
Glutamic acid	10.55	12.29	9.7		
Tyrosine	3.46	4.04	3.5	2.4	2.8
Lysine	6.42	6.90	6.0	4.8	5.4
Histidine	2.31	2.15	1.9	1.3	2.0
Arginine	5.38	4.95	5.1	4.2	4.1
Cystine	0.68	1.48	0.6	0.5	0.6
Tryptophan	1.31	1.40	1.7		

[a] ICI.
[b] Kuraishi, M., et al., *DECHEMA Monogr.*, 83, 111, 1979.
[c] Mogren, H., *Process. Biochem.*, 3, 2, 1979.
[d] Faust, U., *DECHEMA Monogr.*, 83, 125, 1979.

Table 3
CRUDE ANALYSIS OF METHANOL SCP PRODUCTS

	NP[a]	PRUTEEN[b]	MPC (B)[c]	MPC[c]	HOECHST[d]
Dry matter %	95	91		94	
Crude protein (N × 6.25) %	81	74	86	52	82.9
Ash %	7.5	8.2	8.1	7.9	8.5
True protein %	67	62.2	65.2		
Lipid %		8.0	9.6	6.2	

[a] From Mogren, H., *Proc. Biochem.*, 3, 2, 1979.
[b] From ICI.
[c] From Kuraishi, M., et al., *DECHEMA Monogr.*, 83, 111, 1979.
[d] From Faust, U., *DECHEMA Mongr.*, 83, 125, 1979.

REFERENCES

1. **Mogren, H.,** SCP from methanol — the Norprotein process, *Process Biochem.*, 3, 2, 1979.
2. **Faust, U.,** Process results from SCP — pilot plant based on methanol, *DECHEMA Monogr.*, 83, 125, 1979.
3. **Ericsson, M., Ebbinghaus, L., and Lindbloom, M.,** Single cell protein from methanol: economic aspects of the Norprotein process, *J. Chem. Tech. Biotechnol.*, 31, 33, 1981.

4. **Kuraishi, M., Tereo, H., Ohkouchi, N., Matsuda, N., and Nagai, I.,** SCP process development with methanol as substrate, *DECHEMA Monogr.*, 83, 111, 1979.
5. **Hamer, G.,** Biomass from natural gas, in *Microbial Biomass Economic Microbiology,* Vol. 4, Rose, A. H., Ed., Academic Press, New York, 1979.
6. **Romantschuk, H.,** The Pekilo process: protein from spent sulphur liquor, in *Single Cell Protein II,* Tannenbaum, S. R. and Wang, D. I. C., Eds., MIT Press, Cambridge, 1974.
7. **Laine, B. and Chauffaut, J.,** *Single Cell Protein II,* Vol. 4, Tannenbaum, S. R. and Wang, D. I. C., Eds. MIT Press, Cambridge, 1974.
8. **Anthony, C.,** *The Biochemistry of Methylotrophs,* 1982.
9. **Van Dijken, J. P., Oostra-Demkes, G. J., Otto, R., and Harder, W.,** S-formylglutathione: the substrate for formate dehydrogenase in methanol utilising yeasts, *Arch. Microbiol.*, 111, 77, 1976.
10. **Van Dijken, J. P., Otto, R., and Harder, W.,** Oxidation of methanol and formate by catalase purified from methanol grown *Hansenula polymorpha, Arch. Microbiol.*, 106, 221, 1975.
11. **Gow, J. S., Littlehailes, J. D., Smith, S. R. L., and Walter, R. B.,** SCP production from methanol: bacteria, in *Single Cell Protein II,* Tannenbaum, S. R. and Wang, D. I. C., Eds., MIT Press, Cambridge, 1974.
12. **Faust, U. and Sittig, W.,** SCP from methanol: the Hoechst process, in *27th International Congress IUPAC,* Pergamon Press, Oxford, 1980.
13. **Stringer, D.,** Industrial development and evaluation of new protein sources, *Proc. Nutr. Soc.*, 41, 289, 1982.

Chapter 6

OTHER APPLIED ASPECTS OF METHYLOTROPHS

Ching T. Hou

TABLE OF CONTENTS

I. INTRODUCTION

For many years, the primary industrial application for methylotrophs has been the production of single-cell protein, and development in this area continues. Chapter 5 presents a thorough review of this area. Another area, reviewed by Moringa and Hirose in Chapter 4, is the production of secondary metabolites, particularly amino acids. Other areas of application for methylotrophs have been suggested, unfortunately, with no follow-up studies. Silverman[1] suggested the application of methane-oxidizing bacteria to the walls or roofs or coal mines for removal of methane. The analyses of methanotrophs in oil-bearing soil have been suggested for geomicrobiological prospecting of petroleum.[2,3] However, methane-oxidizing bacteria are widespread in nature and methane measured in soils that are not associated with petroleum is in the order of 5 to 15 ppm. Thus, the presence of methane-oxidizing bacteria is not always indicative of the presence of petroleum or natural gas.[4] A microbial fuel cell has also been suggested, using methylotrophs and substrate as an anode, and a molecular oxygen electrode as a cathode.[1]

Recently, the discovery of the extensive range of chemical transformations catalyzed by methanotrophs has opened up the possibility of their exploitation as biocatalysis.[5-7] Since then, many papers and several patents have been published on this subject. It seems likely that such processes will be developed within the next few years. This review covers the accumulated information regarding the use of methylotrophs and/or their enzymes for the transformation or production of petrochemicals. A bioengineering evaluation of an epoxidation process is also included. In addition, patent updates covering all aspects of methylotrophs starting from 1970 to the present are listed.

II. PRODUCTION OF EPOXIDES

Epoxides are extremely valuable industrial chemicals because of their ability to polymerize under thermal, ionic, and free radical catalysis to form epoxy homopolymers and copolymers. Ethylene oxide and propylene oxide constitute the two most important commercial epoxides. Over 90% of the current industrial output of propylene oxide is produced by either the chlorohydrin or the oxirane processes. Both of these processes require multiple steps and depend heavily on the market price of their by-products, such as styrene.

Microbial epoxidation of simple olefins was first demonstrated by Van der Linden,[8] who reported the production of 1 mg of 1,2-epoxyoctane in 30 min at 30°C from 1-octene by 52 mg of heptane-grown cells of *Pseudomonas aeruginosa* resuspended in 40 mℓ of buffer. Further studies on the epoxidation of 1-octene were provided by Cardini and Jurtshuk,[9] who demonstrated that a cell-free extract of a *Corynebacterium* sp. oxidized 1-octene to 1,2-epoxyoctane in addition to hydroxylation octane to octanol. Subsequently, Abbott and Hou[10] demonstrated the epoxidation of 1-octene by whole cells of *Pseudomonas oleovorans*. They found that the methyl group of the product 1,2-epoxyoctane was susceptible to the hydroxylation reaction. All of the epoxidation systems mentioned so far, however, were found to be not active on gaseous alkenes. More recently, Hou et al.[6] discovered that resting-cell suspensions of methylotrophic bacteria epoxidized gaseous alkenes and hydroxylated gaseous alkanes.

A. Whole Cells Systems

Resting-cell suspensions of methane-grown cells of all strains tested oxidized gaseous 1-alkenes to their corresponding 1,2-epoxides, which accumulated (Table 1). The product epoxide accumulated extracellularly, an advantage for industrial production operation. Representatives of three distinct groups of methylotrophic bacteria, *Methylosinus trichosporium* OB3b (an obligate type II methanotroph), *Methylococcus capsulatus* CRL M1(an obligate

Table 1
EPOXIDATION OF PROPYLENE TO PROPYLENE OXIDE BY METHANE-GROWN MICROBES

Methylotrophs	Epoxidation rate (μmol/h/mg of protein)
Obligate, type II	
Methylosinus sp.(CRL 15)	2.2
Methylosinus trichosporium (OB3b)	1.8
Methylosinus sp. (CRL 16)	1.6
Methylosinus sporium (5)	1.0
Methylocystis sp. (CRL 18)	0.7
Methylocystis parvus (OBBP)	0.8
Obligate, type I	
Methylomonas sp. (CRL 4)	1.0
Methylomonas sp. (CRL 21)	1.2
Methylomonas methanica (S_1)	1.3
Methylomonas sp. (CRL 8)	0.6
Methylomonas albus (BG 8)	0.7
Methylomonas sp. (CRL 17)	0.5
Methylomonas sp. (CRL 22)	0.8
Methylomonas sp.(CRL M6P)	0.5
Methylomonas sp. (CRL 20)	1.0
Methylomonas sp. (CRL 10)	0.7
Methylobacter sp. (CRL M6)	0.6
Methylobacter sp. (CRL 23)	1.2
Methylobacter sp. (CRL M1Y)	0.7
Methylobacter sp. (CRL 19)	1.1
Methylobacter sp. (CRL 5)	0.8
Methylococcus capsulatus (CRL M1)	2.5
Methylococcus capsulatus (Y)	0.7
Methylococcus sp. (CRL 25)	0.6
Methylococcus sp. (CRL 24)	0.9
Facultative	
Methylobacterium organophilum (CRL 26)	1.3
Methylobacterium organophilum (XX)	0.9

From Hou et al., *Appl. Environ. Microbiol.*, 38, 127, 1979. With permission.

type I methanotroph), and *Methylobacterium sp.* CRL 26 (a facultative methanotroph), were studied in further detail. In time course studies of epoxide formation, it was found that the rate of propylene oxide production was linear for the first 120 min for strains CRL M1 and OB3b, and for the first 60 min for strains CRL 26 (Figure 1). The reaction, it seems, was limited by the depletion of cofactor, $NADH_2$. The product epoxide was not further metabolized by these methanotrophs. The optimum pH and temperature for epoxide production by all three strains was between 6.0 and 7.0 and around 35°C, respectively. At 40°C, there was an apparent decrease in the amount of epoxide accumulated. This is possibly a result both of the instability of the monooxygenase system at higher temperature and of the volatility of the product propylene oxide (bp 35°C). A propylene concentration in the gaseous phase of approximately 15% supported maximum propylene oxide production. Higher propylene concentrations in the gaseous phase did not stimulate or inhibit the production of propylene oxide. The reaction rate appears to be dependent upon the solubility of both substrates, propylene and oxygen. The substrate specificity for the epoxidation of 1-alkenes by the

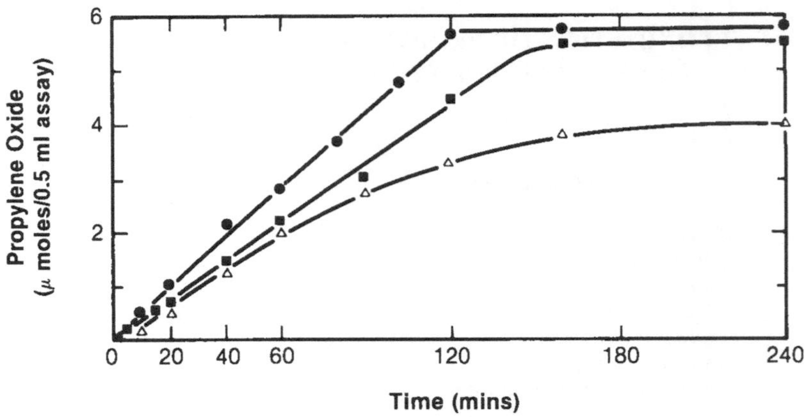

FIGURE 1. Time course of propylene oxide production by resting-cell suspensions of methylotrophic bacteria. *Methylosinus trichosporium* OB3b (●), *Methylococcus capsulatus* CRL M1(■); *Methylobacterium organophilum* CRL 26 (△). (From Hou et al., *Appl. Environ. Microbiol.*, 38, 127, 1979. With permission.)

methanotrophs was determined. The highest production of epoxides was for propylene in all three strains (Table 2). Branched-chain alkenes were also epoxidized. Methane competes with propylene in the oxidation reaction by the whole cell systems.

B. Bioengineering Evaluation

Preliminary bioengineering evaluation of the bioepoxidation process has been conducted. Based on 1983 prices, the prices for propylene oxide and propylene were 45¢/lb and 20¢/lb, respectively. There is a maximum of about 25¢/lb allowance to spend by this epoxidation process in order to compete with the known chlorohydrin/oxirane processes. It is nearly impossible for any biological process to operate at this kind of margin. However, if the production of biocatalyst is also aimed for single-cell protein, then the biocatalyst production cost can be reduced. The spent biocatalyst can be used as animal feed. Methanol is likely to be preferred over methane as a feedstock to produce the biocatalyst. Cloned cells, containing the relevant enzymes, will give a wider choice of feedstock. Nevertheless, the following points remain critical:

Catalyst life time — Whole cells of methylotrophs showed monooxygenase activity during a 12-hr continuous operation.[11] How to keep the reaction rate constant for a longer period of time, for example, one month or half a year, is vitally important for this bioepoxidation process.

Regeneration of the biocatalyst — The bioepoxidation requires a cofactor, reduced NAD, for the reaction. In a whole cell system, the future of this process depends upon development of a way to regenerate the oxidized cofactor back to its reduced form. There have been reports dealing with this topic, e.g., NAD was reduced back to NADH by dehydrogenases and their substrates (see Section V).

Conversion rate — The biocatalyst conversion rate has to be two orders of magnitude greater than those cited in the reports in order to be industrially attractive.

Cooling cost — Cooling cost can become a major concern in an exothermic process operating at 30°C (about 130,000 Btu/lb mol theoretical heat of reaction for propylene oxide, if propylene to oxygen consumption ratio is one). The use of thermophilic methylotrophs may eliminate this problem.

Process pressure to achieve mass transfer — A major rate-limiting step will be substrate

Table 2
OXIDATION OF ALKENES AND METHANE BY RESTING-CELL SI ENSIONS OF METHANE-GROWN BACTERIA

	Methane to methanol	Ethylene to ethylene oxide	Propylene to propylene oxide	1-Bu t 1,2-e but	Butadiene to 1,2-epoxy butene	Isobutene to 1,2-epoxy isobutane
Obligate						
Type II membrane structure						
Methylosinus trichosporium OB3b	1.6	1.9	3.6	0.	2.5	0.32
Type I membrane structure						
Methylococcus capsulatus CRL M1 NRRL B11219	2.5	5.5	5.5	1.	4.4	0.46
Facultative						
Methylobacterium organophilum						
CRL 26 NRRL B-11222	0.7	0.9	2.5	0.	2.8	0.25

Note: In µmole products/hr/2 mg cell (dry weight).

(oxygen and propylene) solubility in the liquid in which the biocatalyst operates. The solubility of oxygen is lower than that of propylene (5.8×10^{-4} g mol/ℓ vs. 27.8×10^{-4} g mol/ℓ at 1 atm pressure). It may be helpful to carry out the reaction under pressure, e.g., 5 to 10 atmospheres, to raise the partial pressure of the reactants.

Genetic engineering certainly could lead to improvements in some problems mentioned above. However, there is no guarantee that the bioepoxidation process will replace the conventional one in the future. Another approach in studying methylotrophs (methane monooxygenase) is to elucidate the enzyme active site configuration, and try mimicking the enzyme active site.

C. Heterogeneous Bioreactor

Attempts were made for the production of propylene oxide using an immobilized biocatalyst (cells) in a continuous mode. Since the substrate is in a gaseous state, it is crucial to have the substrate reach cells and diffuse through the cell membrane freely. Too much alteration of the cell membrane will result in difficulty in diffusing the gaseous substrate through these membranes. Covalent attachment of cells of methanotrophs onto water-insoluble polymers or glass bead resulted in loss of monooxygenase activity. Polyacrylamide-trapped cells of methanotrophs also showed no epoxidation activity.

After many attempts, we successfully ran a heterogeneous bioreactor for the continuous production of propylene oxide.[11] We found that the epoxidation activity of methanotrophic bacteria could be retained when the cells were physically adsorbed to the surface of glass beads. Glass beads which had been coated with cells of methanotrophs were then packed in a glass column. The column (bioreactor) was maintained at 40°C which is above the boiling point of the product, propylene oxide (35°C). The mixture of gaseous substrates (a mixture of propylene and oxygen 1:1, v/v) was introduced through a water bottle *6* (which was also maintained at 40°C) to pick up moisture and then into the bottom of the gas-solid bioreactor *1* (Figure 2). The relative humidity inside the bioreactor was maintained at about 70%. The product propylene oxide was recovered by cooling using ice as the coolant *4*. This heterogeneous bioreactor produced propylene oxide linearly for 7 hr at a rate of 18 μmol/hr/mg protein (Figure 3). After 7 hr of continuous operation, the rate of propylene oxide production slowed down, possibly due to the depletion of endogeneous reducing power (cofactor, NADH). After 10 hr of operation, propylene oxide production essentially stopped.

In situ regeneration of the biocatalyst in the bioreactor was conducted after 12 hr of operation using methanol as the regeneration substrate. The substrate gas mixture inside the jar *8* was replaced with air. The air was forced to pass through a methanol bottle *7* (maintained at 40°C) and then into the bioreactor, carrying methanol vapor with it. The regeneration of biocatalyst continued for 30 min. At the end of the regeneration, the substrate mixture (propylene and oxygen) was reintroduced. The production of propylene oxide immediately resumed at a constant rate of 12 μmol/hr. After an additional 6 hr of operation, the reaction rate slowed down again, indicating the need for additional cofactor regeneration.

The spent biocatalyst (cells) can be easily recovered from the glass beads, e. g., by water-washing, for use as single-cell protein. Although this process requires further improvement, it does point out a potential approach to the application of biocatalysts in the production of commodity chemicals from gaseous hydrocarbons.

III. PRODUCTION OF ALCOHOLS FROM *n*-ALKANES

The oxidation of *n*-alkanes by resting-cell suspensions of methanotrophs produces both primary and secondary alcohols. Both of these alcohols are further oxidized. Primary alcohol is oxidized rapidly into aldehyde and acid, and was not detectable in an assay system using whole cells.[19] Secondary alcohol is further oxidized to methyl ketone (see Section IV). The

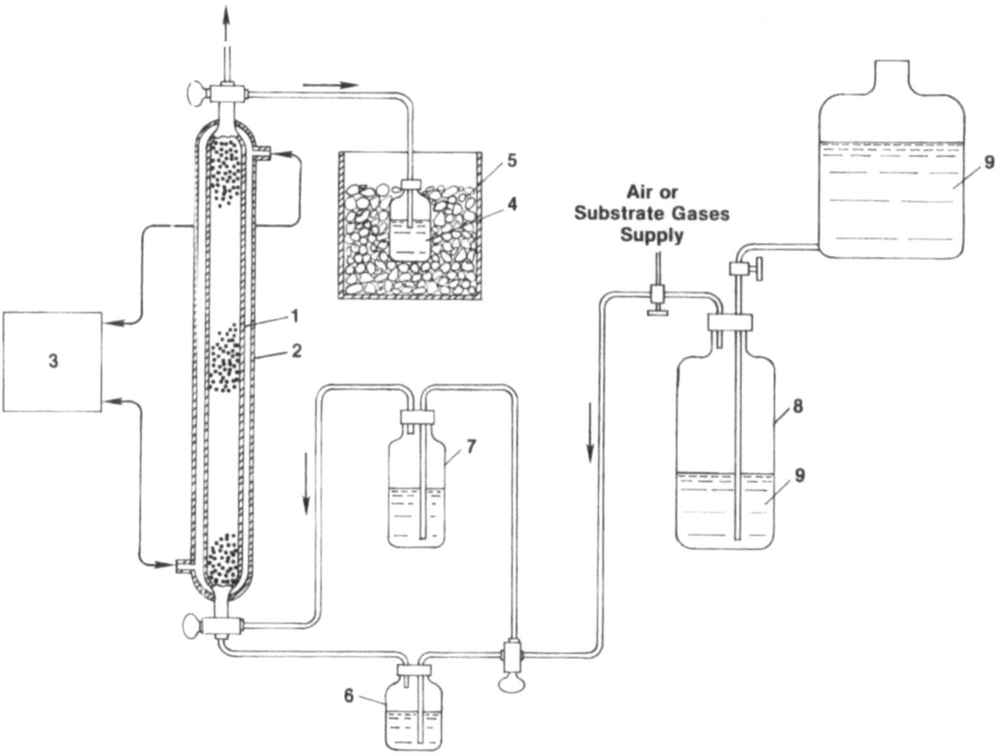

FIGURE 2. Schematic diagram of the gas/solid bioreactor system with biocatalyst regeneration capacity: (1) bioreactor; (2) jacket; (3) temperature control device; (4) liquid for product recovery; (5) ice; (6) water at 40°C; (7) methanol for biocatalyst regeneration; (8) gaseous substrate mixture; (9) water.

production of secondary alcohol from alkane was monitored along with the production of methyl ketone (Figure 4). Since these alcohols did not accumulate, it is not a promising system for microbial production of alcohols at an industrial scale. Among alcohols, methanol has the highest industrial potential of being produced from methane by microbiological system.

Methane is the main ingredient of natural gas produced in huge volume in oil fields. Methane can be liquified and stored at −162°C under atmospheric pressure. Natural gas is transported either by pipelines or as liquified natural gas by refrigerated tankers. However, storage and transportation of both are expensive and hazardous.[12] Methane can be converted to methanol and can be transported and stored as easily as gasoline and should be cheaper and less risky to handle than natural gas or liquified natural gas. Currently, there is no direct chemical process for the conversion of methane to methanol. Current technology in the synthesis of methanol from methane involves the following three steps.[13]

$$CH_4 + H_2O \xrightarrow[\text{Ni-catalyst}]{700\text{-}900°C} CO + 3H_2 \qquad \Delta H_{800} = +226.9\text{KJ/mol}$$
$$\text{15-20\% on } Al_2O_3 \text{ or } SiO_2 \qquad (1)$$

$$CO_2 + H_2 \xrightarrow{\text{Ni catalyst}} CO + H_2O \qquad \Delta H_{25} = +38.5\text{KJ/mol} \qquad (2)$$

$$CO + 2H_2O \xrightarrow[\text{Cu catalyst}]{230\text{-}280°} CH_3OH \qquad \Delta H_{25} = -90.9\text{KJ/mol} \qquad (3)$$

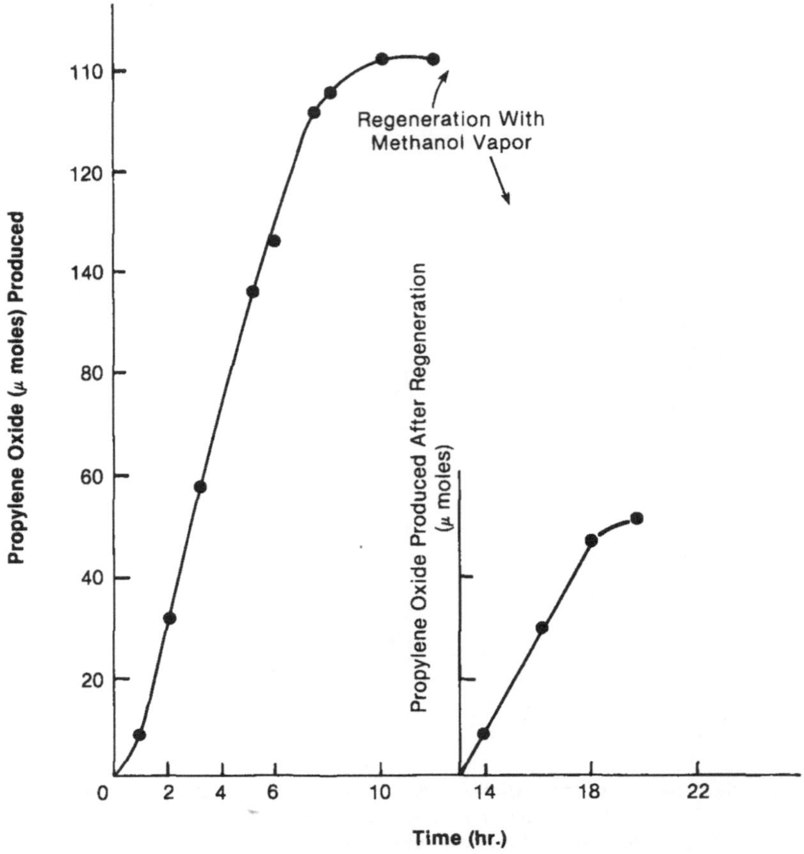

FIGURE 3. Production of propylene oxide from propylene by cells of methanotrophs packed in a gas/solid bioreactor and the regeneration of the biocatalyst.

The steam reforming of methane (Equation 1) yields a synthesis gas rich in hydrogen. The addition of CO_2 shifts the CO/H_2 ratio to an ideal one (1:2) for the next reaction (Equation 2). Finally, the synthesis of methanol from the synthesis gas is an exothermic reaction (Equation 3). All of these steps are energy intensive and involve expensive catalyst.

Microbial oxidation of methane yields methanol. However, methanol is further metabolized instantly without accumulation. In order to produce methanol from methane, the methanol dehydrogenase activity of the cells should be stopped. Tonge et al.[14] found that high phosphate concentration inhibited methanol dehydrogenase activity, resulting in the accumulation of methanol. Better enzyme inhibitors or genetic engineered strains which lack methanol dehydrogenase activity might be an approach to this microbial process for the production of methanol from methane.

IV. PRODUCTION OF METHYL KETONES

Methyl ketones are important industrial solvents. They are currently produced by two processes: (1) two-step process, and (2) direct oxidation process.

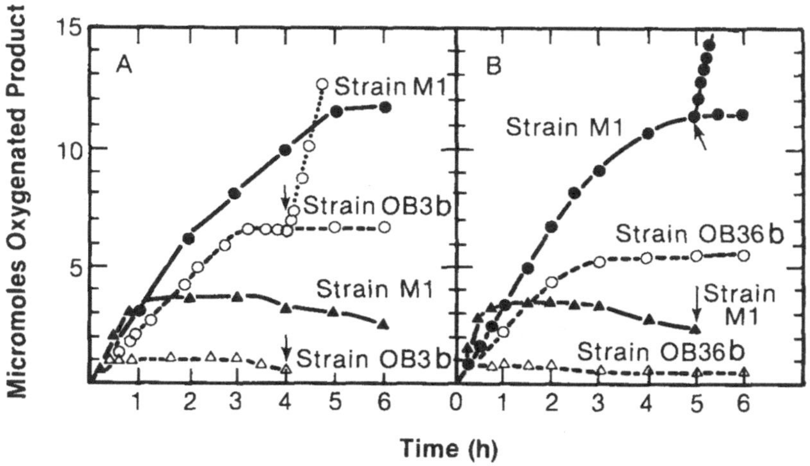

FIGURE 4. Time courses for the production of methyl ketones and secondary alcohols from gaseous alkanes. Cell (solid line for *Methylococcus capsulatus* CRL M1, broken line for *Methylosinus trichosporium* OB3b) were incubated with (A) propane and (B) *n*-butane at 30°C. Products acetone or 2-butanone (○ and ●) and 2-propanol or 2-butanol (△ and ▲) were detected by GLC. Arrow points indicate the time at which external secondary alcohol (2-propanol or 2-butanol) was added. (From Hou et al., *Can. J. Microbiol.*, 27, 107, 1981. With permission.)

(I) Two step process:

(a) Hydration:

$$\text{Alkene} + H_2O \xrightarrow{H_2SO_4} \text{sec-alcohol}$$

(b) Dehydrogenation:

$$\text{sec-alcohol} \longrightarrow \text{methyl ketone} + H_2$$

(II) Direct oxidation process:

$$\text{Alkene} + 2CuCl_2 + H_2O \rightarrow \text{methyl ketone} + 2CuCl + 2HCl$$

$$2CuCl + 2HCl + {}^1/_2O_2 \rightarrow 2CuCl_2 + H_2O$$

The two-step process requires lower capital investment and results in a lower production cost for methyl ketone than does the direct oxidation process. The major reason is that in the direct oxidation process, the aqueous catalyst solution is very acidic and corrosive, and the equipment that contacts it is made of costly corrosion-resistant materials such as titanium.

Methyl ketone formation is well known in mammals and in fungi. In these cases, the ketone is formed by decarboxylation of a β-keto acid and has, therefore, one less carbon atom than the precursor. In contrast, bacterial formation of methyl ketones from alkanes is a unique α-oxidation, with no change in the carbon skeleton.

The first bacterial ketone formation from gaseous alkanes was demonstrated by Leadbetter and Foster[15] with methane-grown *Pseudomonas methanica*. Propane and butane were oxidized but not assimilated by this strain in the presence of the growth substrate (methane). Products of this cooxidation were *n*-propanol, propionic acid, and acetone from propane; and *n*-butanol, butyric acid, and 2-butanone from butane. Subsequently, Lukins and Foster[16] reported that propane-grown *Myocbacterium smegmatis* produced much more methyl ketones and less neutral volatile substances calculated as *n*-propanol. Recently, we found that resting-cell suspensions of methane-grown methanotrophs oxidized *n*-propane and *n*-butane to their

Table 3
OXIDATION OF *n*-ALKANES TO METHYLKETONES BY CELL SUSPENSIONS OF METHANE-UTILIZING ORGANISMS GROWN ON METHANE

Microorganism	Conversion rate (μmol/hr/5 mg of protein)	
	n-Propane to acetone	*n*-Butane to 2-butanone
Known strains		
Methylosinus trichosporium (OB3b)	1.5	1.2
Methylosinus sporium (5)	1.1	0.58
Methylocystis parvus (OBBP)	1.8	0.45
Methylomonas methanica (S_1)	1.5	0.30
Methylomonas albus (BG8)	1.7	0.60
Methylobacter capsulatus (Y)	1.6	1.1
Methylococcus capsulatus (Texas)	1.2	0.52
Methylobacterium organophilum (XX)	1.8	1.0
Newly isolated strains		
Methylosinus sp. (CRL-15)	1.5	1.2
Methylomonas methanica (CRL-21)	1.1	1.5
Methylobacterium sp. (CRL-26)	1.4	1.0
Methylobacter sp. (CRL-M6)	1.0	1.0
Methylobacter bovis (CRL-M1Y)	0.5	0.90
Methylococcus capsulatus (CRL-M1)	1.4	2.0

From Patel et al., *Appl. Environ. Microbiol.*, 39, 727, 1980. With permission.

corresponding methyl ketones.[17,18] In addition, we found that methyl ketones were produced more efficiently from secondary alcohols by resting cell suspensions of methanotrophs.[17]

A. From *n*-Alkanes

Resting cell suspension of methanotrophs oxidized *n*-alkanes (C_3 to C_6) to their corresponding methyl ketones[18,19] (Table 3). Among *n*-alkanes, propane and *n*-butane were oxidized faster. Secondary alcohols were detected as intermediates during oxidation of *n*-alkanes (Figure 4). The product methyl ketones were not metabolized further and accumulated extracellularly. Branch-chain alkanes such as isobutane and isopentane were also oxidized. The optimum pH and temperature for the production of methyl ketones from *n*-alkanes by resting cell suspensions of methanotrophs were 6.0 to 7.0 and 30 to 35°, respectively. The reaction involves two oxidation steps. The first step is the oxidation of *n*-alkanes to secondary alcohols catalyzed by methane monooxygenase. The second step is the oxidation of secondary alcohols to methyl ketones catalyzed by a novel enzyme, secondary alcohol-specific alcohol dehydrogenase.[20-24] The rate-limiting factor in the production of methyl ketones from *n*-alkanes was found to be the first step, the oxidation of *n*-alkanes to secondary alcohols.[19]

In the first step, primary alcohols are also produced along with secondary alcohols, from *n*-alkanes. Primary alcohols are oxidized further to aldehydes and acids. This affects the yield of methyl ketones from *n*-alkanes. The multiple product formation in this process limits its possible industrial application.

B. From Secondary Alcohols

We first demonstrated the production of methyl ketones from secondary alcohols by

Table 4
PRODUCTION OF 2-BUTANONE FROM 2-BUTANOL BY CELL SUSPENSIONS OF C_1-UTILIZING MICROBES

Microbes	Conversion rate (μmol/hr/mg of protein)
Obligate methane utilizers (methane-grown)	
Type II membrane structures	
Methylosinus trichosporium (OB3b)	4.8
Methylosinus sp. (CRL 15)	4.5
Methylosinus sporium (5)	3.0
Methylosinus sp. (CRL 16)	2.5
Methylocystis parvus (OBBP)	1.1
Methylocystis sp. (CRL 18)	1.0
Type I membrane structure	
Methylomonas methanica (S_1)	0.4
Methylomonas albus (BG8)	3.5
Methylomonas sp. (CRL 4)	2.5
Methylomonas sp. (CRL 8)	3.0
Methylomonas sp. (CRL 10)	0.5
Methylomonas sp. (CRL 17)	2.0
Methylomonas sp. (CRL 20)	2.0
Methylomonas sp. (CRL 21)	2.8
Methylomonas sp. (CRL 22)	1.5
Methylomonas sp. (CRL M6P)	1.4
Methylomonas sp. (CRL M6)	1.4
Methylomonas sp. (CRL 23)	1.0
Methylomonas sp. (CRL M1Y)	3.0
Methylomonas sp. (CRL 19)	1.8
Methylomonas sp. (CRL 5)	2.0
Methylococcus capsulatus (CRL M1)	5.0
M. capsulatus (Y)	0.8
Methylococcus sp. (CRL 25)	0.9
Methylococcus sp. (CRL 24)	2.5
Facultative methane utilizers (methane-grown)	
Methylobacterium organophilum (CRL 26)	2.5
Methylobacterium organophilum XX	2.5
Obligate methanol utilizers	
Pseudomonas sp. CRL 75 ATCC 21439	5.4
Methylomonas methylovora ATCC 21852	2.5
Facultative methanol utilizers (methanol-grown)	
Pseudomonas sp. CRL 74 ATCC 21438	3.2
Pseudomonas Ms. ATCC 25262	3.5
Yeasts (methanol-grown)	
Candida boidinii NRRL Y-2332	6.0
Candida utilis ATCC 26387	6.8
Hansenula polymorpha ATCC 26012	5.8
Hansenula anomala NRRL Y-336	5.5
Pichia pastoris NRRL Y-55	5.8
Pichia sp. CRL 72	4.0

From Hou et al., *Appl. Environ. Microbiol.*, 3, 135, 1979. With permission.

resting-cell suspensions of methane- and methanol-grown microbes.[17] The product methyl ketones were not metabolized further and accumulated extracellularly. Table 4 showed the activity for converting 2-propanol and 2-butanol to acetone and 2-butanone, respectively,

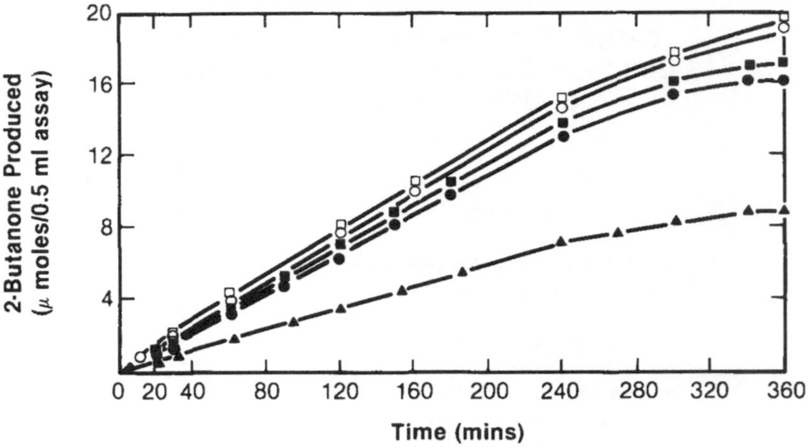

FIGURE 5. Time course of 2-butanone production by resting cell suspensions of methanol-grown cultures from five distinct types of methylotrophs. Resting-cell suspensions (0.7 mg protein in 0.5 mℓ) were incubated with 10 μℓ of 2-butanol at 30°C for various times. Product 2-butanone was assayed by gas chromatography. Symbols: ● *Methylosinus trichosporium* OB3b; ■ *Methylococcus capsulatus* CRL M1; ▲ *Methylobacterium* sp. CRL 26; ○ *Pseudomonas* sp. ATCC 21439; and □ *Hansenula polymorpha* ATCC 26012. (From Hou et al., *Appl. Environ. Microbiol.*, 38, 135, 1979. With permission.)

by C_1-utilizing microbes. All the C_1-utilizers tested were able to produce methyl ketones from secondary alcohols.

Optimum conditions for the production of 2-butanone were compared among five distinct types of methane- or methanol utilizers: *Methylosinus trichosporium* OB3b (type II, obligate), *Methylococcus capsulatus* CRLM1 (type I, obligate), *Methylobacterium* sp. CRL 26 (facultative), *Hansenula polymorpha* ATCC 26012 (yeast), and *Pseudomonas* sp. ATCC 21439 (obligate methanol utilizer).

The production of 2-butanone from 2-butanol reached a maximum after 14 hr of incubation in batch experiments with all the microbes tested. The amount of product 2-butanone had not declined after 30 hr of incubation. The rate of 2-butanone production was linear for the first 4 hr (Figure 5). The optimum pH and temperature for the production of methyl ketones were around 8.0 and 35°C, respectively. The amount of 2-butanone produced was dependent on the amount of substrate initially added; 0.1 *M* 2-butanol supported maximum production. There was no product inhibition in the production of 2-butanone. Among secondary alcohols, 2-propanol and 2-butanol were oxidized at higher rates (Table 5). The (−) enantiomer of 2-butanol was oxidized preferentially over its (+) form.

The production of methyl ketones from secondary alcohols is a rather simple one-step reaction. The rate of the reaction is about 5 to 6 μmol/hr/mg protein. However, after 4 hr of reaction, the system requires resupply of the cofactor, NAD, to maintain its reaction rate. This can also be achieved by coupling this system with another system which recycles $NADH_2$ back to NAD, such as the epoxidation system (see Section II.3).

The system for the production of methyl ketones from secondary alcohols has the potential for further industrial exploitation. The production of stereospecific secondary alcohol from either *n*-alkane or a racemic mixture of secondary alcohol may be attractive industrially.

V. REGENERATION OF COFACTOR

In a cell-free system, the epoxidation of alkenes and the hydroxylation of alkanes catalyzed

Table 5
OXIDATION OF SECONDARY ALCOHOLS BY CELL SUSPENSIONS OF C_1-UTILIZERS GROWN ON METHANOL

	Oxidation rate (µmol/hr/mg of protein)			
C_1-utilizers	2-Propanol to acetone	2-Butanol to 2-butanone	2-Pentanol to 2-pentanone	2-Hexanol to 2-hexanone
Methylosinus trichosporium OB3b	0.30	4.8	2.7	0.09
Methylococcus capsulatus CRL M1	2.0	5.0	0.24	0.08
Methylobacterium organophilum CRL 26	0.72	2.5	1.0	0.09
Hansenula polymorpha ATCC 26012	5.9	5.8	1.4	0.72
Pseudomonas sp. ATCC 21439	3.5	5.4	0.05	0.03

From Hou et al., *Appl. Environ. Microbiol.*, 38, 135, 1979. With permission.

by methane monooxygenase require a cofactor, $NADH_2$. In a whole cell system, the cells can catalyze the methane monooxygenase reaction using endogeneous stored reducing power (cofactor). However, after 1 to 2 hr of reaction, the stored reduced cofactor becomes depleted. The addition of a reduced cofactor, or coupling a cofactor regeneration system with the epoxidation/hydroxylation system becomes necessary. Hou et al.[25] reported the stimulation of propylene epoxidation in whole cell systems of methanotrophs by methane and its metabolites. Both methanol and formate stimulated the production of propylene oxide (Table 6). In cell-free systems, many NAD-linked dehydrogenases were found compatible with the methane monooxygenase system for generation/regeneration of $NADH_2$.[26,27] For example, formate dehydrogenase, formate, and NAD or secondary alcohol dehydrogenase, 2-alcohol, and NAD can generate $NADH_2$ for the methane monooxygenase reaction (Table 7).

No commercial system would appear to be viable using cell extracts because of the instability of the enzyme and the exorbitant price of the electron donor. Recently, there have been successes in making water-soluble macromolecularized NAD which shows activity for several dehydrogenases.[28-30] For the production of high volume commodity chemicals, however, the use of cell-free system is still presently out of the question. The most attractive process may be the combination of producing propylene oxide and methyl ketones using a whole cell system of methanotrophs.

Such a process would consist of two reactions operating simultaneously inside the cells of methanotrophs. The first reaction is the production of propylene oxide from propylene, catalyzed by methane monooxygenase and $NADH_2$. The second reaction involves the production of methyl ketones (such as 2-butanone) from 2-butanol by secondary alcohol dehydrogenase and NAD. Both of the enzymes are known to exist in the cells of methanotrophs. The amount of endogeneous cofactor might be sufficient for the process because the cofactor is regenerated constantly through the second reaction. The process is shown in the following equations:

Table 6
STIMULATION OF EPOXIDATION OF PROPYLENE IN WHOLE CELL SUSPENSIONS OF METHYLOTROPHS BY METHANE METABOLITES

Microbes and metabolites added	Epoxidation rate, propylene oxide (μmol/0.5 mℓ assay[a]) 1 hr	2 hr	3 hr
Methylococcus capsulatus CRL M1			
Control	2.0	3.5	3.6
+ CH_3OH (6 m*M*)	2.7	4.2	5.2
+ CH_3OH (48 m*M*)	1.8	4.2	5.2
+ HCHO (0.25 m*M*)	2.3	3.6	4.5
+ HCHO (4 m*M*)	1.9	3.5	3.6
+ HCHO (10 m*M*)	1.8	3.0	3.5
+ HCOOH (10 m*M*)	2.6	4.0	4.5
+ HCOOH (40 m*M*)	2.1	3.5	4.1
Methylosinus trichosporium OB3b			
Control	2.0	3.0	3.1
+ CH_3OH (6 m*M*)	2.8	4.0	5.1
+ CH_3OH (48 m*M*)	1.9	3.8	5.0
+ HCHO (0.25 m*M*)	2.4	3.4	3.9
+ HCHO (4 m*M*)	2.0	3.0	3.1
+ HCHO (10 m*M*)	1.8	2.4	2.5
+ HCHO (50 m*M*)	0.8	1.2	1.3
+ HCOOH (10 m*M*)	2.7	3.8	4.2
+ HCOOH (40 m*M*)	2.1	3.3	4.1
Methylobacterium organophilum CRL 26			
Control	1.4	2.2	2.3
+ CH_3OH (6 m*M*)	1.8	2.5	3.2
+ CH_3OH (45 m*M*)	1.3	2.5	3.3
+ HCHO (0.25 m*M*)	1.6	2.5	3.0
+ HCH (4 m*M*)	1.2	2.0	2.1
+ HCHO (10 m*M*)	1.2	1.9	2.0
+ HCOOH (10 m*M*)	1.8	2.6	3.2
+ HCOOH (40 m*M*)	1.3	2.4	3.2

[a] Protein contents of each assay were: 0.74 mg (strain CRLMI); 1.1 mg (strain OB3b); and 1 mg (strain CRL 26).

From Hou, et al., *FEMS Microbiol. Lett.*, 9, 267, 1980. with permission.

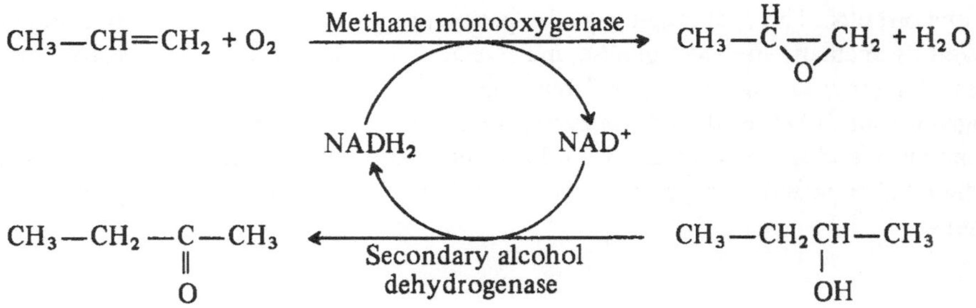

Exogeneous addition of NAD may increase the reaction rate. The process can be operated in a continuous mode by adding both substrates, propylene and 2-butanol, and harvesting both products, propylene oxide and 2-butanone, continuously.

Table 7
EPOXIDATION OF PROPYLENE BY SOLUBLE METHANE MONOOXYGENASE FROM *METHYLOSINUS* SP. CRL 31 USING COFACTOR GENERATED BY VARIOUS DEHYDROGENASES AND THEIR SUBSTRATES

Cofactor generation system	Rate of epoxidation of propylene (nmol/min/mg protein)
Formate dehydrogenase, formate and NAD^+	70
Diol dehydrogenase, 1,2-propanediol and NAD^+	60
Secondary alcohol dehydrogenase, 2-propanol and NAD^+	60
Primary alcohol dehydrogenase, ethanol and NAD^+	75
Control 1 NADH	75
Control 2 NAD^{+2}	0

Note: The specific activities for dehydrogenases added or in the soluble fraction (in nmol/min/mg protein) were: formate dehydrogenase 60; diol dehydrogenase 40; secondary alcohol dehydrogenase 40; and primary alcohol dehydrogenase 80. Substrate for the dehydrogenase and NAD^+ added were 10 μmol and 5 μmol, respectively.

From Hou et al., *J. Appl. Biochem.*, 4, 379, 1982. With permission.

VI. PRODUCTION OF POLY β-HYDROXYBUTYRATE

Poly β-hydroxybutyrate (PBH) is an aliphatic thermoplastic polyester with several properties similar to polypropylene. PHB is biodegradable, nontoxic, and benign to tissue. It is a strong candidate as a surgical material. In addition, its unusual electrical properties gives it a good chance to become commercially viable.

PHB was first described in 1926 by Lemoigne.[31] There have been reports concerning the wide spread occurrence of PHB in bacteria.[32-35] Kallio and co-worker[36,37] first showed at a methylotroph, *Methylomonas methanica,* deposited a large quantity of PHB in cells (about 25% of the dry weight of 4-day-old cultures). A convenient spectrophotometric assay method for PHB was established by Law and Slepecky in 1961.[38] In 1970, Whittenbury et al. reported that many methane-utilizers were packed with lipid inclusions, mainly PHB.[39,40] Two vibrio-shaped methane-oxidizing bacteria, *M. soehngenii*[41] and thermophilic and thermotolerant methylotrophs[42] were found to possess various amounts of PHB. The production of PHB was also noted in the type II methanotroph, *Methylosinus trichosporium* OB3b.[43-45] Thompson et al.[44] found β-hydroxybutyrate dehydrogenase, an enzyme involved in PHB mobilization, in extracts of methylotrophs. They also demonstrated the accumulation of acetone concomitant with PHB degradation. β-Hydroxybutyrate dehydrogenase was partially purified from methane-grown *Methylococcus capsulatus* CRL M1 (Hou[46]). The oxidation of β-hydroxybutyrate to acetoacetate requires a cofactor NAD in its oxidized form. This reaction produces reducing equivalents, NADH, for cells to perform the vitally needed oxygenation reaction catalyzed by methane monooxygenase. The precise reasons for this mobilization of PHB reserves and the seemingly wasteful excretion of acetone under these conditions are not clear. More extensive studies of PHB as to its role as a storage polymer in methanotrophs, the regulation of its synthesis, and the subsequent motilization are needed in order to develop an industrial process for the production of PHB.

VII. APPLICATION IN POLLUTION CLEAN-UP

The ability of methane monooxygenase to oxidize a wide variety of organic compounds including alkanes, alkenes, halogenated hydrocarbons, cyclic and aromatic compounds suggests that methylotrophs can be used in the clean-up of organic pollutants. Their ability to oxidize halogenated C compounds, e.g., chloromethane, trichloromethane, bromomethane, etc., is particularly attractive, since these compounds are inert to attack by most other microbes.

VIII. PATENTS

Patents on methane- or methanol-utilizing microorganisms have been searched covering the period 1970 to the present. The sources used in this study included the comprehensive data base on U.S. Patents, the Derwent Patent File, and the API Patent File. There are a large number of patents on the production of protein. However, we list here only those patents dealing with specialty chemicals. These patents have been arranged in the following groups, according to their application: (1) biotransformation and fine chemicals; (2) production of polysaccharides; and (3) production of amino acids.

Patents in each group are listed in filing date order. The following symbols are used: (a) patent number; (b) filing date; (c) issued or published date; (d) title of patent; and (e) patent assignee.

A. Biotransformation and Fine Chemicals

1. Microbial Oxidations

1. (a) U.S. 4,353,987; (b) June 11, 1981; (c) October 12, 1982; (d) Glyceraldehyde production from glycerol using methanol dehydrogenase especially in or from *Methylobacterium organophilum* NRRL B-12486; (e) Upjohn Co.
2. (a) U.S. 4,348,476; (b) January 22, 1981; (c) September 7, 1982; (d) Production of epoxides such as propylene oxide using packed catalytic bed containing moist resting cells exhibiting oxygenase activity; (e) Exxon Res. and Eng. Co.
3. (a) G.E. 2,081,306; (b) August 1, 1980; (c) February 17, 1982; (d) Microbiological oxidation of alkane(s) and alkene(s) with methane-utilizing bacteria adapted to utilize liquid methanol; (e) Imperial Chem. Inds. Ltd.
4. (a) U.S. 4,368,267; (b) February 6, 1980; (c) January 11, 1983; (d) Epoxidation of lower α-olefins; (e) Exxon Res. and Eng. Co.
5. (a) E.P. 19,937; (b) June 5, 1979; (c) December 10, 1980; (d) Preparation of alcohol oxidase solutions from methanol-using *Pichia*-type microorganisms by homogenization, and removal of solids; (e) Phillips Petroleum Co.
6. (a) U.S. 4,268,630; (b) May 30, 1979; (c) May 19, 1981; (d) Microbiological production of ketones from C_3-C_6 alkanes; (e) Exxon Res. and Eng. Co.
7. (a) U.S. 4,266,034; (b) March 27, 1979; (c) May 5, 1981; (d) Method for producing microbial cells and use thereof to produce oxidation products; (e) Exxon Res. and Eng. Co.
8. (a) D.T. 2,830,327; (b) July 10, 1978; (c) January 24, 1980; (d) Alcohol oxidase preparation by *Hansenula polymorpha* yeast fermentation — in methanol-containing nutrient solution admixed with nutrient solution containing glycerol, sorbitol or xylose; (e) Kernforchungs Julich.
9. (a) B.E. 884,004; (b) June 29, 1979; (c) October 16, 1980; (d) Enzymatic epoxidation of propylene by hydrogen peroxide — prepared by enzymatic oxidation of methanol with formaldehyde as byproduct; (e) Standard Oil Co. (Ind.).

10. (a) G.B. 2,024,205; (b) May 16, 1978; (c) January 9, 1980; (d) Microbiological oxidation of organic compounds with methane-utilizing *Methylosinus trichosporium* bacteria; (e) Imperial Chem. Inds. Ltd.
11. (a) B.E. 875,512; (b) April 14, 1978; (c) October 12, 1979; (d) Methyl ketone production from *n*-alkane, secondary alcohol by fermentation with preactivated methylotrophic microorganism; (e) Exxon Res. and Eng. Co.
12. (a) U.S. 4,347,319; (b) April 14, 1978; (c) August 31, 1982; (d) Microbiological epoxidation of olefin compounds with methylotrophic bacteria conditioned to utilizing methane; (e) Exxon Res. and Eng. Co.
13. (a) U.S. 4,269,940; (b) April 14, 1978; (c) May 26, 1981; (d) Microbial oxidation of alkane(s) to alcohol(s) and ketone(s) by using enzymes from methylotrophic bacteria and yeasts; (e) Exxon Res. and Eng. Co.
14. (a) U.S. 4,250,259; (b) April 14, 1978; (c) February 10, 1981; (d) Conversion of linear sec-alcohol(s) to methyl ketone(s) by contacting with dehydrogenase from methylotrophic microorganism; (e) Exxon Res. and Eng. Co.
15. (a) U.S. 4,241,184; (b) April 14, 1978; (c) December 23, 1980; (d) Methyl ketone(s) preparation from 3-6 C sec-alcohol(s) using methylotrophic microorganism or secondary alcohol dehydrogenase derived from microorganism; (e) Exxon Res. and Eng. Co.
16. (a) J. 54-017184; (b) July 4, 1979; (c) February 8, 1979; (d) Hydrocarbon oxidation in culture medium by oxidizing alkane, alkene, or organic cyclic compounds using methane-oxidizing bacteria; (e) National Res. Dev. Corp.

2. Wastewater Treatment

1. (a) D.T. 3,088,665; (b) March 9, 1979; (c) October 9, 1980; (d) Biological oxidation of halogen-containing cyclic organic compound, degradation with methane-consuming bacteria or methane monooxygenase and/or dehalogenase; (e) Imperial Chem. Inds. Ltd.
2. (a) U.S. 3,928,191. (b) October 6, 1972; (c) December 23, 1975; (d) Methanolic waste water biodegradation — accelerated with dispersed particulate attapulgite/montmorillonite; (e) Du Pont De Nemours Co.

3. Miscellaneous Chemicals

1. (a) J. 57-033599; (b) August 6, 1980; (c) February 23, 1982; (d) Coenzyme α10 preparation by cultivating microorganism of the genus *Methylobacterium* in medium containing carbon source, vitamin(s), inorganic compounds, etc.; (e) Mitsubishi Gas Chem. Ind.
2. (a) D.T. 3,020,851; (b) June 4, 1979; (c) December 11, 1980; (d) Adenosine 5′-triphosphate microbiological production by fermentation of an ATP-producing, methanol assimilating bacterium in a medium containing methanol and inorganic phosphate; (e) Kureha Kagaku Kogyo K.K.
3. (a) J. 55-096091; (b) January 17, 1979; (c) July 21, 1980; (d) Ray fungi production by culturing vitamin B12 producing fungi of genus *Pseudonocardia* in methanol-containing medium with addition of cobalt source; (e) Kawamura Inst. Chem. Res.
4. (a) J. 55-061796; (b) October 31, 1978; (c) May 9, 1980; (d) High linoleic acid content microbial oil and fat production by culturing candida microorganism using methanol as carbon source; (e) Agency of Ind. Sci. Tech.
5. (a) J. 55-042534; (b) September 19, 1978; (c) March 25, 1980; (d) D-Amino: acylase manufactured from methanol-assimilating *Pseudomonas* bacteria — to give enzyme that specifically hydrolyzes acyl group n *N*-acyl-2-amino acids used in penicillin and cephalosporin derivative preparation; (e) Sanraku Ocean.

6. (a) J. 55-019022; (b) July 25, 1978; (c) February 9, 1980; (d) Microbiological quantitative determination of vitamin B12 by cultivating vitamin B12-requiring microorganism which assimilates methanol and determining growth of microorganism; (e) Sanraku Ocean.
7. (a) J. 54-080495; (b) February 27, 1978; (c) September 11, 1979; (d) Coenzyme α8 production by culturing suitable methanol-utilizing bacteria of *Pseudomonas* or *Methylomonas* genus; (e) Mitsubishi Gas Chem. Ind.
8. (a) J. 54-080495; (b) December 8, 1977; (c) June 27, 1979; (d) Riboflavin production by culturing yeast of genus *Hansenula* in culture medium containing methanol; (e) Mitsubishi Gas Chem. Ind.
9. (a) J. 52-090692; (b) January 22, 1976; (c) July 30, 1977; (d) Coenzyme α7 prepared from yeast of genus *Hansenula* by culturing in medium containing methanol as carbon source; (e) Mitsubishi Toatsu Chem. Inc.
10. (a) U.S. 3,809,615; (b) June 6, 1969; (c) June 7, 1974; (d) Methane detecting system — based on enzymes from microorganism active for methane oxidation; (e) D. L. Klass.

B. Production of Polysaccharides

1. (a) J. 56-140896; (b) April 2, 1980; (c) November 4, 1981; (d) Fermentative preparation of polysaccharide(s) using *Pseudomonas* microorganisms in culture medium containing methanol; (e) Mitsubishi Petroch. K.K.
2. (a) J. 56-121497; (b) February 27, 1980; (c) September 24, 1981; (d) Polysaccharide production by cultivating microorganism of genus *Hyphomicrobium* in culture medium containing methanol, nitrogen source, inorganic salt, and other additives; (e) Japan Tobacco and Salt Pub.
3. (a) J. 56-099795; (b) January 6, 1980; (c) August 11, 1981; (d) Polysaccharide production by cultivating *Pseudomonas* microorganisms in medium containing methanol as main carbon source; (e) Mitsubishi Petrochem K.K.
4. (a) J. 56-061996; (b) October 22, 1979; (c) May 27, 1981; (d) Polysaccharide microbiological production — using a *Pseudomonas* microorganism, and methanol as the carbon source, used as an anticancer agent; (e) Mitsubishi Petrochem K.K.
5. (a) J. 56-6048891; (b) September 26, 1979; (c) May 2, 1981; (d) Production of polysaccharide by cultivation of *Pseudomonas* useful as anticancer, cholesterol-lowering or antiviral agent, or food additive. (e) Mitsubishi Yuka K.K.
6. (a) J. 56-001887; (b) June 20, 1979; (c) January 10, 1981; (d) Coenzyme α10 production from *Acremonium* or *Paecilomyces* spp. by fermentation; (e) Jujio Paper Mfg. K.K.
7. (a) U.S. 4,261,420; (b) April 30, 1979; (c) April 14, 1981; (d) Carbon dioxide assisted oil recovery — using gas from integrated single-cell protein plant; (e) Provesta Corp.
8. (a) J. 55-120601; (b) March 9, 1979; (c) September 17, 1980; (d) Polysaccharide MH-2 production by culturing suitable methane-utilizing bacteria; (e) Ajinomoto K.K.
9. (a) B.E. 866934; (b) May 11, 1977; (c) November 10, 1978; (d) Polysaccharide preparation from *Pseudomonas polysaccharogenes* — useful for reducing blood levels of cholesterol; (e) Mitsubishi Petroch. K.K.
10. (a) B.E. 857493; (b) August 6, 1976; (c) February 6, 1978; (d) Selecting bacterial strains to produce polyhydroxy butyric acid which is used to give moldings etc. which are biodegradable; (e) Agroferm AG.
11. (a) B.E. 857486; (b) August 4, 1976; (c) February 6, 1978; (d) Fermentative production of D-(3)-hydroxybutyric acid — useful in parenteral feeding and as intermediate for perfumes; (e) Agroferm A.G.

12. (a) U.S. 4,006,058; (b) November 24, 1975; (c) February 1, 1977; (d) Heteropolysaccharide production by fermentation of methanol in a medium containing alkali metal glycerophosphate, has improved viscosity kinetics and yield; (e) Mobile Oil Corp.
13. (a) J. 51-054987; (b) November 7, 1974; (c) May 14, 1976; (d) Polysaccharides preparation using novel strain — *Methanomonas* genus cultured in methanol medium; (e) Mitsubishi Gas Chem. Ind.
14. (a) U.S. 3,878,045; (b) June 12, 1973; (c) April 15, 1975; (d) Novel heteropolysaccharides prepared by fermentation using methanol as sole source of assimilable carbon; (e) Cornell Res. Foundation.
15. (a) Fr. 2231748; (b) May 30, 1973; (c) January 31, 1975; (d) Extracellular polysaccharides preparation by cultivating bacteria capable of converting methanol as carbon source; (e) Inst. Francais Du Petrole.
16. (a) U.S. 4,061,585; (b) May 29, 1973; (c) December 6, 1977; (d) Methods of using a bacterial heteropolysaccharide for dispersing, flocculation, reducing frictional drag, emulsifying, and preventing oil deposition; (e) Cornell Res. Foundation.
17. (a) U.S. 3,923,782; (b) May 29, 1973; (c) December 2, 1975; (d) Heteropolysaccharide by culturing *Methylomonas mucosa* on a medium containing methanol as sole source of assimilable carbon; (e) Cornell Res. Foundation.
18. (a) U.S. 4,932,218; (b) May 29, 1973; (c) January 13, 1976; (d) Heteropolysaccharide production by fermenting *Methylomonas mucosa* on a methanol-containing medium useful esp. in secondary oil recovery; (e) Cornell Res. Foundation.
19. (a) U.S. 4,016,085; (b) May 29, 1973; (c) April 5, 1977; (d) Heteropolysaccharides from methanol using *Methylomonas* — useful as thickening agents in flooding compositions; (e) Cornell Res. Foundation.
20. (a) U.S. 3,965,985; (b) August 12, 1968; (c) June 29, 1976; (d) Microbial synthesis from oxygenated hydrocarbon feedstock giving polymeric gum production — useful as water flood additive, drilling mud additive, and plugging agent; (e) Phillips Petroleum Co.

C. Production of Amino Acids

1. (a) J. 57-110193; (b) December 29, 1980; (c) July 8, 1982; (d) L-Methionine preparation by fermentation of *Pseudomonas* spp. mutant which is resistant to ethionine and alpha-amino beta-hydroxy-valeric acid; (e) Ajinomoto K.K.
2. (a) J. 57-115189; (b) December 29, 1980; (c) July 17, 1982; (d) L-Methionine preparation useful as amino acid transfusion or feed additive by reacting methanol or betaine and homocysteine with *Pseudomonas, Microcyclus* or *Methylomonas* microorganism; (e) Ajinomoto K.K.
3. (a) J. 7-063096; (b) September 30, 1980; (c) April 16, 1982; (d) L-Methionine preparation from *Methylomonas* genus bacteria includes forming the dissociant OEA-9 which resists ethionine and alpha-amino-beta-hydroxy valeric acid; (e) Ajinomoto K.K.
4. (a) J. 56-127095; (b) March 12, 1980; (c) October 5, 1981; (d) Leucine preparation by fermentation, e.g., of microorganism of *Corynebacterium* genus in presence of L-valine, carbon and nitrogen sources; (e) Kyowa Hakko Kogyo K.K.
5. (a) J. 56-035992; (b) August 29, 1979; (c) April 8, 1981; (d) Preparation of L-methionine by fermentation of *Escherichia coli* strain which is resistant to growth suppression by lysine, phenylalanine or histidine analogues; (e) Ajinomoto K.K.
6. (a) J. 53-034987; (b) September 9, 1976; (c) March 31, 1978; (d) L-leucine or L-valine preparation by culturing a microorganism of genus *Methylomonas* using methanol as carbon source; (e) Tani, Y.
7. (a) J. 53-009393; (b) July 14, 1976; (c) January 27, 1978; (d) Glutathione used as cosmetic additive and in hepatic treatment is prepared by culturing *Methylomonas* and

Pseudomonas in medium containing methanol or methane and sepg. glutathione; (e) Kanebo K.K.

8. (a) D.T. 2554530; (b) December 4, 1975; (c) June 16, 1977; (d) Microbiological production of L-serine by aerobic cultivation of *Pseudomonas* DSM 672 or 673 on a medium containing methanol and glycine; (e) Ges Biotechno.
9. (a) D.T. 2508515; (b) March 5, 1974; (e) September 11, 1975; (d) L-Tryptophan production by fermentation — using methanol and methanol-consuming bacteria which react with glycine to form serine; (e) Bofors AB.
10. (a) J. 50-042094; (b) August 16, 1973; (c) April 16, 1975; (d) Preparing amino acids, organic acids, and vitamins by cultivating *Hyphomicrobium* microorganisms in methanol medium; (e) Ajinomoto K.K.
11. (a) D.T. 2438206; (b) August 11, 1973; (c) February 20, 1975; (d) L-Glutamic acid preparation by bacterial fermentation in methanol-containing nutrient medium using *Pseudomonas* or *Protaminobacter* mutants; (e) Kyowa Hakko Kogyo Co. Ltd.
12. (a) J. 50-025792; (b) July 16, 1973; (c) March 18, 1975; (d) L-Tyrosine and L-phenylalanine production by *Methanomonas* or *Protaminobacter* culture using methanol-containing nutrient medium; (e) Kyowa Termentation K.K.
13. (a) D.T. 2417337; (b) April 10, 1973; (c) October 31, 1974; (d) Biotechnical production of L-lysine by culturing methanol-utilizing *Protaminobacter* mutants; (e) Kyowa Hakko Kogyo Co.
14. (a) U.S. 3856774; (b) August 12, 1968; (c) December 24, 1974; (d) Microbial synthesis of proteins, gums, amino acids, etc. from aldehyde containing hydrocarbon derived products; (e) Phillips Petroleum Co.
15. (a) Fr. 20051860; (b) March 30, 1968; (c) December 5, 1969; (d) Fermentation method to produce ℓ-glutamic acid; (e) Ajinomoto Co., Inc. and Sanraku Ocean Ltd.
16. (a) U.S. 3,222,258; (b) July 15, 1963; (c) N.A.; (d) Method of preparing amino acids by fermentation; (e) Ajinomoto Co. Inc.

REFERENCES

1. **Silverman, M. P.,** Inform. Circ. 8246, *U.S. Bur. Mines*, 1964, 1.
2. **Strawinski, R. J.,** Prospecting, U.S. Patent 2665237, 1954.
3. **Davis, J. B.,** Geomicrobial Prospecting Method for Petroleum, U.S. Patent 2777799, 1957.
4. **Davis, J. B. and Updegraff, D. M.,** Microbiology in the Petroleum Industry, *Bacteriol. Rev.*, 18 (4) 215, 1954.
5. **Colby, J., Stirling, D. I., and Dalton, H.,** The Soluble methane monooxygenase of *Methylococcus capsulatus* (Bath). Its ability to oxygenate n-alkanes, n-alkenes, ethers, and alicyclic, aromatic and heterocyclic compounds, *Biochem. J.*, 165, 395, 1977.
6. **Hou, C. T., Patel, R. N., Laskin, A. I., and Barnabe, N.,** Microbial oxidation of gaseous hydrocarbons: epoxidation of C_2 to C_4 n-alkenes by methylotrophic bacteria, *Appl. Environ. Microbiol.*, 38, 127, 1979.
7. **Higgins, I. J., Hammond, R. C., Saviaslani, F. S., Best, D., Davies, M. M., Tryhorn, S. E., and Taylor, F.,** Biotransformation of hydrocarbons and related compounds by whole organism suspension of methane-grown *Methylosinus trichosporium* OB3b, *Biochem. Biophys. Res. Commun.*, 89, 671, 1979.
8. **Van der Linden, A. C.,** Epoxidation of α-olefins by heptane-grown *Pseudomonas* cells, *Biochim. Biophys. Acta*, 77, 157, 1963.
9. **Cardini, G. and Jurtshuk, P.,** Cytochrome P-450 involvement in the oxidation of n-octane by cell-free extracts of *Corynebacterim* sp. strain 7E1C, *J. Biol. Chem.*, 245, 2789, 1970.
10. **Abbott, B. J. and Hou, C. T.,** Oxidation of 1-alkenes to 1,2-epoxy-alkanes by *Pseudomonas oleovorans*, *Appl. Microbiol.*, 26, 36, 1973.

11. **Hou, C. T.**, Production of epoxides such as propylene oxide using packed catalytic bed containing moist resting cells exhibiting oxygenase activity, U.S. Patent 4,348,476, 1982.
12. **Davis, L. N.**, Gambling on frozen fire, *New Sci.*, 85, 70, 1980.
13. **Ghisalba, O. and Heinzer, F.**, Methanol from methane — a hypothetical microbial conversion compared with the chemical process, *Experientia*, 38, 218, 1982.
14. **Tonge, G. M., Harrison, D. E. F., Knowles, C. J., and Higgins, I. J.**, Properties and partial purification of methane-oxidizing enzyme system from *Methylosinus trichosporium*, *FEBS Lett.*, 58, 293, 1975.
15. **Leadbetter, E. R. and Foster, J. W.**, Bacterial oxidation of gaseous alkanes, *Arch. Mikrobiol.*, 35, 92, 1960.
16. **Lukins, H. B. and Foster, J. W.**, Methyl ketone metabolism in hydrocarbon-utilizing *Mycobacteria*, *J. Bacteriol.*, 85, 1074, 1963.
17. **Hou, C. T., Patel, R. N., Laskin, A. I., Barnabe, N., and Marczak, I.**, Microbial oxidation of gaseous hydrocarbons: production of methyl ketones from their corresponding secondary alcohols by methane- and methanol-grown microbes, *Appl. Environ. Microbiol.*, 38, 135, 1979.
18. **Patel, R. N., Hou, C. T., Laskin, A. I., Felix, A., and Derelanko, P.**, Microbial oxidation of gaseous hydrocarbons: production of methyl ketones from corresponding n-alkanes by methane-utilizing bacteria, *Appl. Environ. Microbiol.*, 39, 727, 1980.
19. **Hou, C. T., Patel, R. N., Laskin, A. I., Marczak, I., and Barnabe, N.**, Microbial oxidation of gaseous hydrocarbons: production of alcohols and methyl ketones from their corresponding n-alkanes by methylotrophic bacteria, *Can. J. Microbiol.*, 27, 107, 1981.
20. **Hou, C. T., Patel, R. N., Laskin, A. I., Barnabe, N., and Marczak, I.**, Identification and purification of nicotinamide adenine dinucleotide dependent secondary alcohol dehydrogenase from C-1 utilizing microbes, *FEBS Lett.*, 101, 179, 1979.
21. **Hou, C. T., Patel, R. N., Barnabe, N., and Marczak, I.**, Stereospecificity and other properties of a novel secondary alcohol-specific alcohol dehydrogenase, *Eur. J. Biochem.*, 119, 395, 1981.
22. **Patel, R. N., Hou, C. T., Laskin, A. I., Derelanko, P., and Felix, A.**, Microbial production of methylketones: purification and properties of a secondary alcohol dehydrogenase from yeast, *Eur. J. Biochem.*, 101, 401, 1979.
23. **Patel, R. N., Hou, C. T., Laskin, A. I., and Derelanko, P.**, Microbial production of methyl ketones; properties of a purified yeast secondary alcohol dehydrogenase, *J. Appl. Biochem.*, 3, 218, 1981.
24. **Hou, C. T., Patel, R. N., Laskin, A. I., Barnabe, N., and Marczak, I.**, Substrate specificity and stereospecificity of NAD-linked alcohol dehydrogenase from methanol-grown yeasts, *Appl. Environ. Microbiol.*, 41, 829, 1981.
25. **Hou, C. T., Patel, R. N., Laskin, A. I., and Barnabe, N.**, Microbial oxidation of gaseous hydrocarbons: oxidation of lower *n*-alkenes and n-alkanes by resting cell suspensions of various methylotrophic bacteria, and the effect of methane metabolites, *FEMS Microbiol. Lett.*, 9, 267, 1980.
26. **Patel, R. N., Hou, C. T., Laskin, A. I., and Felix, A.**, Microbial oxidation of hydrocarbons: properties of a soluble methane monooxygenase from a facultative methane-utilizing organism, *Methylobacterium* sp. CRL 26, *Appl. Environ. Microbiol.*, 44, 1130, 1982.
27. **Hou, C. T., Patel, R. N., Laskin, A. I., and Barnabe, N.**, Epoxidation of alkenes by methane monooxygenase: generation and regeneration of cofactor, $NADH_2$, by dehydrogenases, *J. Appl. Biochem.*, 4, 379, 1982.
28. **Fuller, C. W., Rubin, J. R., and Bright, H. J.**, A simple procedure for a covalent immobilization of NADH in a soluble and enzymatically active form, *Eur. J. Biochem.*, 103, 421, 1980.
29. **Okada, H.**, Synthesis of macromoleculized-NAD^+ and its application, *Hakkokogaku Kaishi*, 56, 441, 1978.
30. **Muramatsu, M., Urabe, I., Yamada, Y., and Okada, H.**, Synthesis and kinetic properties of a new NAD^+ derivative carrying a vinyl group, *Eur. J. Biochem.*, 80, 111, 1977.
31. **Lemoigne, M.**, Produits de deshydration et de polymerization de l'acid β-oxybutyrique, *Bull. Soc. Chim. Biol.*, 8, 770, 1926.
32. **Forsyth, W. G. C., Haward, A. C., and Roberts, J. B.**, Occurrence of PHB in aerobic Gram-negative bacteria, *Nature (London)*, 182, 800, 1958.
33. **Haward, A. C., Forsyth, W. A. C., and Roberts, J. B.**, Synthesis and breakdown of poly-β-hydroxybutyric acid by bacteria, *J. Gen. Microbiol.*, 20, 510, 1959.
34. **Morris, M. B. and Roberts, J. B.**, A group of Pseudomonads able to synthesize poly-β-hydroxybutyric acid, *Nature (London)*, 183, 1538, 1959.
35. **Levine, H. B. and Wolochow, H.**, Occurrence of PHB in *Pseudomonas pseudomallei*, *J. Bacteriol.*, 79, 35, 1960.
36. **Harrington, A. A. and Kallio, R. E.**, Oxidation of methanol and formaldehyde by *Pseudomonas methanica*, *Can. J. Microbiol.*, 6, 1, 1960.
37. **Kallio, R. E. and Harrington, A. A.**, Sudanophilic granules and lipids of *Pseudomonas methanica*, *J. Bacteriol.*, 80, 321, 1960.

38. **Law, J. H. and Slepecky, R. A.,** Assay of poly-β-hydroxybutyric acid, *J. Bacteriol.*, 82, 33, 1961.
39. **Whittenbury, R., Phillips, K. C., and Wilkinson, J. F.,** Enrichment, isolation and some properties of methane-utilizing bacteria, *J. Gen. Microbiol.*, 61, 205, 1970.
40. **Whittenbury, R., Davies, S. L., and Davey, J. F.,** Exospores and cysts formed by methane-utilizing bacteria, *J. Gen. Microbol.*, 61, 219, 1970.
41. **Hazeu, W. and Steenis, P. J.,** Isolation and characterization of two vibrio shaped methane oxidizing bacteria, *Antonie van Leeuwenhoek, J. Microbiol. Serol.*, 36, 67, 1970.
42. **Malashenko, Y. R.,** Isolation and characterization of new species (thermophilic and thermotolerant ones) of methane-utilizers, in *Symp. Microbial Production Utilization Gases (H_2, CH_4, CO)*, Schlegel, H. G., Gottschalk, G., and Pfennig, N., Ed., Gottingen Akademie der Wissenschaften, Gottingen, 293, 1976.
43. **Weaver, T. L., Patrick, M. A., and Dugan, P. R.,** Whole-cell and membrane lipids of the methylotrophic bacterium *Methylosinus trichosporium, J. Bacteriol.*, 124, 602, 1975.
44. **Thompson, A. W., O'Neill, J. G., and Wilkinson, J. F.,** Acetone production by methylobacteria, *Arch. Mikrobiol.*, 109, 243, 1976.
45. **Best, D. J. and Higgins, I. J.,** Methane-oxidizing activity and membrane morphology in a methanol-grown obligate methanotroph, *Methylosinus trichosporium,* OB36, *J. Gen. Microbiol.*, 125, 73, 1981.
46. **Hou,** unpublished data.

INDEX

A

B

C

D

E

F

G

H

I

K

N

P

Q

R

S

T

U

V

W

X

Y